Germán Tomas
Adrián Acuña

Oil biomarkers

Germán Tomas
Adrián Acuña

Oil biomarkers

Characterization and environmental stability of biomarkers in oil from the Austral Basin, Santa Cruz Province

ScienciaScripts

Imprint

Any brand names and product names mentioned in this book are subject to trademark, brand or patent protection and are trademarks or registered trademarks of their respective holders. The use of brand names, product names, common names, trade names, product descriptions etc. even without a particular marking in this work is in no way to be construed to mean that such names may be regarded as unrestricted in respect of trademark and brand protection legislation and could thus be used by anyone.

Cover image: www.ingimage.com

This book is a translation from the original published under ISBN 978-613-8-98229-6.

Publisher:
Sciencia Scripts
is a trademark of
Dodo Books Indian Ocean Ltd. and OmniScriptum S.R.L publishing group

120 High Road, East Finchley, London, N2 9ED, United Kingdom
Str. Armeneasca 28/1, office 1, Chisinau MD-2012, Republic of Moldova, Europe
Printed at: see last page
ISBN: 978-620-7-92415-8

Contents

ACKNOWLEDGEMENTS

To my Director, Dr. Adrián Acuña "General Coordinator of the Regional Laboratory of Forensic Investigation of the Judicial Power of the Province of Santa Cruz" for trusting me to carry out this research, for his time, patience, collaboration, coordination in the techniques and tests used throughout this time, for being my mentor and guide in this training path.

To the Universidad Tecnológica Nacional - Facultad Regional Santa Cruz for being my second home during these five years in which I developed my doctoral training in such a prestigious institution. To the Dean, Mr. Sebastián Puig, for supporting the research projects that arose from my doctoral training and for financing through the UTN - FRSC my trips and stays at the numerous congresses I have participated in throughout the country.

To the Universidad Nacional de la Patagonia Austral for giving me the opportunity to carry out my postgraduate studies at their university.

To the Centro de Investigación y Transferencia de la Provincia de Santa Cruz for giving me the opportunity to carry out this research in response to its regional economy.

To CONICET for providing me with grants throughout the development of this doctoral work.

To the Judicial Power of the Province of Santa Cruz for giving me access to the facilities of the Regional Laboratory of Forensic Investigation to carry out the chromatographic analysis of the crude oil samples.

To YPF -Tecnología for allowing me to carry out part of my PhD research by financing all the required analyses.

To Fomicruz Sociedad del Estado for supplying me with crude oil samples from the Del Mosquito oilfield and for allowing me to visit their facilities.

To the Compañía General de Combustibles for being the main supplier of the crude oil samples used in this PhD study.

To my co-director, Dr. Walter Vargas, "Head of the Biotechnology Area of Y-TEC", who welcomed me cordially into his team, passing on his knowledge and making important contributions to my work.

To the Biochemist Liliana Coggiola for accepting me in the Chair of General Chemistry at the Santa Cruz Regional Faculty - National Technological University. For teaching me how to teach in the university environment during the doctoral training process.

To Javier Szewczuk for his kind assistance in obtaining samples and providing relevant information for the development of this study.

To Engineer Mariano Bertinat, "Secretary of State for the Environment of the Province of Santa Cruz", for his collaboration and management in obtaining the sampling permits for the sites studied.

To the engineer Leandro Almonacid for his collaboration in the preparation of the cartography presented both in the doctoral thesis and in all the publications in journals.

To Dr. Marcos E. Escobar Navarro of the University of Zulia, Faculty of Engineering, Maracaibo, Venezuela for his invaluable and disinterested advice in the interpretation of the results obtained. May you rest in peace!

To Dr. Sandra Casas for her advice on each of the concerns raised during my training over the years.

To my daughters ELUNEY and AMISCIA, for accompanying me on this journey, for being a reason not to give up when I wanted to give up, for being everything I can't express in words,

I love you!

To the love of my life Danella that I never forgot and searched for so long, after 16 years I could find you. Because our love remained unchanged through the days and you came at the end of my career to give me that last push, I love you!

To my mum who left before I reached this goal. Wherever you are, thank you for giving me life and protecting me in the face of all the adversities you went through. Thank you for instilling in me the importance of study, its intangibility and value. I will always love you.

To my brothers Adolfo and Diego, for always encouraging me and giving me strength in every decision, even when adversities arose, I love you!

To my aunt Raquel, for all your teachings, and the love you have given me as if I were your son. You have been a fundamental pillar during my personal and academic development, for always motivating me to climb one more step on this infinite ladder of knowledge. I love you.

To my great friends Tomás Contreras and Jesús Ponce for being like brothers to me and although the distance is great, we know that we are close. I love you.

To my other great friend and colleague Enrique Alum, thank you for sharing time with me, those colourful chats, that knowledge and understanding of life that makes you for me a person with an enormous potential to achieve any goal. I love you.

Finally, I dedicate this work to all those people who gave me their unconditional support.

Thank you all!

SUMMARY

Title: "Characterisation and environmental stability of biomarkers in oil from the Austral Basin, Santa Cruz Province".

Oil spills or oil theft can be a problem in the Province of Santa Cruz due to its hydrocarbon-producing nature. This research was focused on evaluating the potential of molecules present in crude oil known as biomarkers to respond to the aforementioned problems. Based on the above, the study was divided into two stages: first, the chemical characterisation by biomarkers of 15 samples of light crude oil from the Springhill Fm and the Lower Magallanes Fm (Southern Basin), which were obtained in quadruplicate over a period of one year. In the second stage, from one of the previously analysed crude oils, an artificial weathering test was carried out under laboratory conditions in both seawater and soil for 12 months. The aim was to evaluate the behaviour of biomarkers against this physical, geochemical and biological process that normally alters the general composition of oil when it is released into the environment. In the two working instances, all crude oil samples were diluted in n-pentane and conditioned with silica gel to generate a colourless extract. It was then subjected to solid-liquid column adsorption chromatography to obtain its aliphatic and aromatic fractions. These organic solutions were then analysed using a gas chromatograph coupled to a mass spectrometer to determine their profiles of n-alkanes, aromatic compounds and biomarkers. The results associated with the chemical characterisation of the 15 crude oils studied suggest that the precursor organic matter was of a marine-continental type, deposited in an environment with moderate oxygen concentration and in which the source rocks (Palermo Aike Fm and Margas Verdes Fm) evidenced a marine nature, mainly associated with silico-clastic sedimentation. However, all the hydrocarbons differed from each other with a decreasing degree of association around the following parameters: oil well > reservoir > puncture depth > geological formation > paleobiodegradation. On the other hand, evaporation and biodegradation were the phenomena that promoted the changes observed in the samples subjected to induced weathering in the laboratory, yet the biomarkers remained stable during the monitoring year. Therefore, the use of biomarkers as a resolution tool in situations involving crude oil spills or the theft, transport and/or storage of stolen hydrocarbons is feasible as they constitute a unique and reliable chemical signature that identifies each crude oil.

Key words: Esteranes, terranes, Springhill Fm, Lower Magallanes Fm, artificial weathering.

1. INTRODUCTION

The following is a review of the theoretical foundations of this research. It begins with the concept of petroleum, then describes the stages of its formation, and how its aliphatic and aromatic fractions are composed. Subsequently, emphasis is placed on the importance of biomarkers and their diagnostic relationships within the oil industry, their behaviour in the face of weathering and the scope of this research. Finally, the general characteristics of the Austral Basin, i.e. its geological setting, are developed.

1.1 Oil.

Oil is a substance of heterogeneous nature consisting of three phases: solid, liquid and gaseous (Speight 2014). It is naturally located beneath the earth's surface as a result of long processes. These processes include its generation from organic matter that settled and then during burial under favourable temperature and pressure conditions, its maturation from hydrocarbon source rocks, its expulsion and migration through faults and permeable rocks, and finally its entrapment in impermeable barriers known as traps (Speight 2014). From the point of view of its elemental composition (Table 1), petroleum consists mainly of C and H. In addition, heteroatoms (S-NO) and metals (V, Ni, among others) combine to form organic molecules as simple as methane to very complex macromolecules such as asphaltenes (Klemt et al. 2020).

Table 1. Chemical composition of crude oil *and* gas. Taken from López *and* Lo Mónaco (2017).

Element	Concentration in crude oil	Concentration in gas
Carbon	84 - 87 %	65 - 80 %
Hydrogen	11 - 14 %	1 - 25 %
Sulphur	0,06 - 2 %	0 - 0,2 %
Nitrogen	0,1 - 2 %	1 - 15 %
Oxygen	0,1 - 2 %	0 %
Vanadium	5 - 1300 ppm	Not applicable
Nickel	1 - 150 ppm	Not applicable

ppm = parts per million.

Crude oil is a complex mixture of hydrocarbons that exists in the liquid phase of oil reservoirs preserved in porous and/or permeable rocks. It maintains its liquid state at atmospheric pressure after coming to the surface, but its chemical composition, viscosity and API (American Petroleum Institute) gravity may vary due to the origin of organic matter in the geological units and basins to which they belong (API 2016).

1.1.1 Diagenesis, catagenesis, and metagenesis.

Hydrocarbon generation is a process that is divided into three stages: diagenesis, catagenesis and metagenesis. Diagenesis is the biological, physical and chemical alteration of organic matter in sediments, prior to significant changes caused by temperatures < 50 °C (Kujawinski et al. 2002; Vargas-Escudero et al. 2021). Early in this stage one of the main agents of transformation of organic matter is microbial activity whereby proteins and polysaccharides are degraded into amino acids and simple sugars, respectively (Kim et al. 2005; Dowey et al. 2020). These monomers can be incorporated into insoluble polymeric structures by a process of polycondensation. These macromolecules are known as humic substances (fulvic acids, humic acids and humins) and account for more than 90 % of the organic matter present in sediments (Truskewycz et al. 2019). Kerogen is the end product of diagenesis and is the material from which liquid and gaseous hydrocarbons are formed (Vandenbroucke and Largeau 2007). There are three types of kerogen: type I (alginite or sapropelic), type II (exinite or humic) and type III (vitrinite). In addition, the idea of a type IV kerogen (inertinite) is raised because, in some cases, characteristics dissimilar to the three previously established types were observed (Table 2).

Table 2. Types of kerogen *and* their characteristics. Taken from Vandenbroucke *and* Largeau 2007.

Types	Content	Precursor organic matter	Generate
I	High in C and H and low in O	Lacustrine, occasionally marine and animal	Light crude oil
II	Moderate in C, H *and* O	Marine and animal	S-rich oil
III	Moderate in C, H *and* O	Non-marine organic matter	Natural Gas
IV	Inert particles	Rusty	Coal

C = carbon, H = hydrogen, O = oxygen, S = sulphur.

Catagenesis is the process by which kerogen is modified (chemically and physically) when subjected to high pressures and temperatures, the latter being the most important factor. Organic matter present in sedimentary rocks is altered in a temperature range that can range from 50 to 150 °C (Vandenbroucke and Largeau 2007; Schito and Corrado 2018). Under typical burial conditions, these transformations occur over millions of years. They are characterised by the breaking of bonds in the kerogen, and the formation of smaller molecules (devoid of oxygen), which will become part of the bitumen (Garcette-Lepecq et al. 2000; Vargas-Escudero et al. 2021). Finally, metagenesis is a process in which residual kerogen and liquid hydrocarbons are cleaved into gas at temperatures between 150 and 200 °C. It is the most advanced stage of bitumen maturation. It is the most advanced stage of maturation of organic matter before it enters metamorphism and thermogenic CH4 is its main product. At this point the conditions are such that no compound can resist them and cleavage of all hydrocarbons occurs (Vandenbroucke and Largeau 2007; Dowey et al. 2020). In other words, these are the processes that take place in a source rock and are summarised in Figure 1.

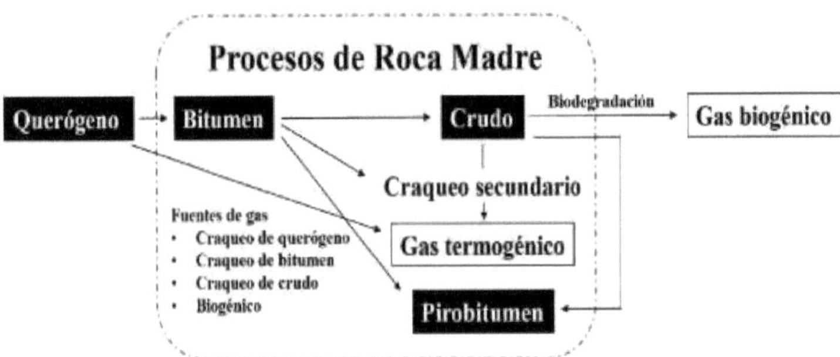

Figure 1. Processes occurring in a source rock. Adapted from Jarvie et al. 2007.

The chemical composition of petroleum varies depending on several factors, such as the organic matter that gave rise to it and the conditions during which it sedimented, the type of source rock that generated it and the thermal regime that prevailed during its formation, among others (López and Lo Mónaco 2017). However, oil can generally be separated into four fractions: aliphatic, aromatic, resin and asphaltenes. It should be noted that there is a limitation to study resins and asphaltenes by gas chromatography because they volatilise above 350 °C, hence only the lighter portions are worked with in research involving biomarkers (Stashenko et al. 2014). The first oil fraction is composed of aliphatic compounds, i.e. without aromaticity, and is the most important in this type of studies due to the number of biomarkers present in it (Killops and Killops 2005; Chen et al. 2018). The aliphatic groups of interest for this Thesis were the n-alkanes and the biomarkers known as pristane, phytane, terpanes and steranes (Peters et al. 2005).

1.1.2.1 n-Alkanes.

N-alkanes are the most abundant hydrocarbons in petroleum and are characterised by an open-chain molecular structure consisting only of carbon and hydrogen (Kuppusamy et al. 2020). They have the potential to provide a wealth of information about the crudes of which they are a part, and their modification by environmental weathering is a key aspect of the nature of these investigations (Khan et al. 2018). All n-alkanes involved in this work are presented below (Table 3).

Table 3. Names, symbols and formulae of the n-alkanes identified.

Name	Symbol	Formula	Name	Symbol	Formula
nonano	n-C9	C9H20	dean	n-C10	C10H22
undean	n-C11	C11H24	dodecano	n-C12	C12H26
tridecane	n-C13	C13H28	tetradecane	n-C14	C14H30
pentadecane	n-C15	C15H32	hexadecane	n-C16	C¼H34
heptadecane	n-C17	C17H36	octadecane	n-C18	C18⅛8
nonadecano	n-C19	C19H40	eicosane	n-C20	C20H42
eneicosan	n-C21	C21H44	docosan	n-C22	C22H46
tricosan	n-C23	C23H48	tetracosane	n-C24	C24H50
pentacosane	n-C25	C25H52	hexacosane	n-C26	C26H54
heptacosane	n-C27	C27H56	octacosan	n-C28	C28H58
nonacosano	n-C29	C29⅛0	triacontano	n-C30	C30⅛2

1.1.2.2 Pristane and phytane.

Biomarkers, initially referred to as biomarkers or geochemical fossils, are organic molecules, mostly derived from the lipids of living organisms (Killops and Killops 2005). These compounds contain information about the environmental conditions under which the organic matter settled and the transformation processes it underwent (Gaines et al. 2009; Orea et al. 2021). He et al. (2018) described biomarkers as complex molecules present in oil that are characterised by high thermal stability during diagenesis and catagenesis. This intrinsic property is associated with the chemical structure of each of the biomarkers, remaining without major structural changes, which allows inferring the biological precursors from which they derive (Killops and Killops 2005; Costa de Sousa et al. 2022). The most studied biomarkers are pristane (P = 2, 6, 10, 14- tetramethylpentadecane) and phytane (F = 2, 6, 10, 14-tetramethylhexadecane), due to the ease with which they are determined by conventional analytical techniques. Pristane is formed by the diagenetic degradation of phytol (a molecule derived from chlorophyll), through an oxidation process, and phytane must be generated through a reduction process, so the P/F ratio gives clues to the oxidoreductive depositional conditions (Didyk et al. 1978; Chen et al. 2018). Furthermore, they are very important in the geochemical interpretation of crude samples because their concentration is the highest among the other acyclic isoprenoid hydrocarbons present in this fraction (Peters et al. 2005). On the other hand, the ratio of pristane/heptadecane (P/n- C17) to phytane/octadecane (F/n-C18) is used to suggest the type of precursor organic matter (Shanmugam 1985; López et al. 2019).

1.1.2.3 Terpanes.

Other biomarkers of interest are the pentacyclic terpanes known as hopanoids, derived from the cell membranes of prokaryotic organisms (El-Sabagh et al. 2018). The biological compounds from which they are derived known as hopanoids are the ones that best retain their structure during diagenesis and catagenesis (Bost et al. 2001; Kao et al. 2018). Among the hopanoids, the 17α, 21 β (H) hopano C30 (Fig. 2) should be noted as it is generally the dominant one in crude samples, and its ratio to gammacerane (G30) defined as (G30/H30) x 100 is an indicator of the salinity present in the medium (Seifert et al. 1984; Han et al. 2019).

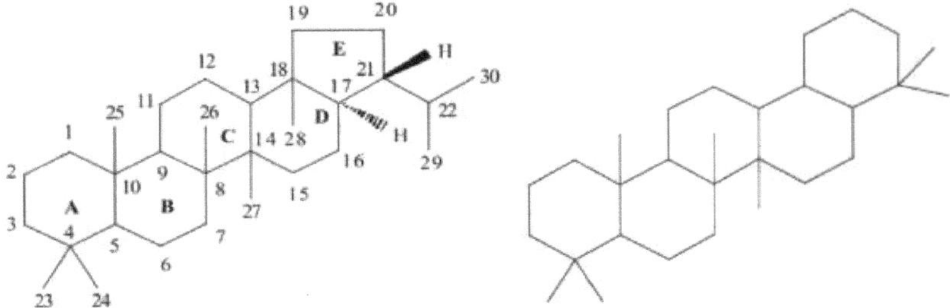

Figure 2. Structures of the pentacyclic terpanes known as hopane C_{30} (left) and gammacerane (right). Taken from Peters et al. 2005.

The tricyclic terpanes are a less numerous but very important group within organic geochemistry because they provide information on the lithology of the source rock, for example, the predominance of the tricyclic terpane C_{23} in the fragmentogram corresponding to ion $m/z = 191$, with respect to terpanes C_{21}, C_{22} and C_{24}, is indicative of the presence of carbonate source rocks (Alberdi et al. 2001; Inglis et al. 2018). The symbology and nomenclature used in the identification of the tri- and pentacyclic terpanes is presented in Table 4.

Table 4. Names, symbols and formulae of identified terpanes.

Name	Symbol	Formula	Name	Symbol	Formula
Tricyclic terpane C_{19}	T19	C19H34	**Trisnorhopane** C_{27}	Ts	C27H46
Tricyclic terpane C_{20}	T20	C20H36	**Trisnorneohopane** C_{28}	Tm	C27H46
Tricyclic terpane C_{21}	T21	C21H38	**Norhopane** C_{29}	H29	C29H50
Tricyclic terpane C_{23}	T23	C23H42	**Hopano** C_{30}	H30	C30H52
Tricyclic terpane C_{24}	T24	C24H44	**Moretano** C_{30}	M30	C30H52
Tricyclic terpane C_{25}	T25	C25H46	**Homohopanos** C_{31}	H31	C31H54
Tricyclic terpane C_{26}	T26	C26H48	**Bishomohopanos** C_{32}	H32	C32H56
Gammacerano C_{30}	G30	C30H52	**Trishomohopano** C_{33}	H33	C33H58

1.1.2.4 Steranes.

Finally, the biomarkers known as steranes (Fig. 3) are naphthenic molecules with a four-ring structure associated with eukaryotic organisms (Rangel et al. 2017). They are found as sterols in algae, animals and higher plants and thus provide information on the precursor organic matter (Killops and Killops 2005). The most commonly used steranes in organic geochemistry contain 27, 28 and 29 carbon atoms and are called cholestanes, ergostanes and stigmastanes, respectively (Peters et al. 2005; Orea et al. 2021). Under certain conditions, steranes undergo transpositions of the methyl groups, catalysed by clay minerals, adopting the most thermodynamically stable configuration, resulting in compounds known as diasteranes (Escobar et al. 2012).

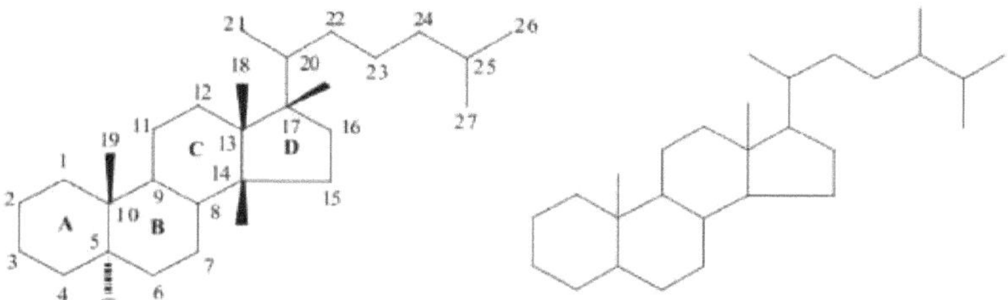

Figure 3. Structures of the steranes known as cholestane (left) and ergostane (right). Taken from Peters et al. 2005.

Pregnanes also belong to this group of biomarkers, but are characterised by a lower number of carbons in the aliphatic chain associated with the tetracycle (Rangel et al. 2017). The symbology and nomenclature used in the identification of pregnanes, steranes and diasteranes is presented in Table 5.

Table 5. Names, symbols and formulae of the steranes identified.

Name	Symbol	Formula	Name	Symbol	Formula
14β-pregnano C20	P20	C21H36	20R-14a-cholestane C27	S27 (aR)	C27H48
14β-pregnano C21	P21	C22H38	20S-14α-ergostane C28	S28 (aS)	C28H50
14β-pregnano C22	P22	C23H40	20R-14β-ergostane C28	S28 (βR)	C28H50
20S-13β-diasterane C27	D27 (βS)	C27H48	20S-14β-ergostane C28	S28 (βS)	C28H50
20R-13β-diasterane C27	D27 (βR)	C27H48	20R-14α-ergostane C28	S28 (aR)	C28H50
20S-13α-diasterane C27	D27 (aS)	C27H48	20S-14α-stygmastane C29	S29 (aS)	C29H52
20R-13α-diasterane C27	D27 (aR)	C27H48	20R-14β-stygmastane C29	S29 (βR)	C29H52
20S-14α-cholestane C27	S27 (aS)	C27H48	20S-14β-stygmastane C29	S29 (βS)	C29H52
20R-14β-cholestane C27	S27 (βR)	C27H48	20R-14α-estigmastanoC29	S29 (aR)	C29H52
20S-14β-cholestane C27	S27 (βS)	C27H48			

1.1.3 Aromatic Fraction.

The second portion of the oil, if we consider increasing polarity as a property, is the aromatic fraction and is made up of a group of molecules that present aromaticity. Among these compounds we can mention polycyclic aromatic hydrocarbons (PAHs) and monoaromatic (MAS) and triaromatic (TAS) biomarkers, the latter being less abundant than their aliphatic peers (Sivan et al. 2008). With respect to PAHs, they are not considered biomarkers because their origin is unspecific, i.e. it is not possible to trace with certainty a palaeobiogeochemical pathway indicating the precursors and/or intermediates until they reach them (Summons and Lincoln 2012). However, they are also a useful tool in this discipline as they provide valuable information. For example, the diagnostic ratio (DR) of dibenzothiophene (DBT) to phenanthrene (Ph) represented as a function of the P/F ratio is used to suggest the lithology of the source rock (Hughes et al. 1995; Walters et al. 2018). In addition, methylphenanthrene is used as a maturity parameter and experimentally correlated with vitrinite reflectance (Ro) to determine the thermal evolution of crude oils (Abdulazeez and Fantke 2017). Therefore, within organic geochemistry these molecules are known as aromatic markers and those that were integrated in this work are presented in Table 6.

Table 6. Names, symbols and formulae of the PAHs identified.

Name	Symbol	Formula	Name	Symbol	Formula
naphthalene	N	C10H8	2,3,6 trimethylnaphthalene	TMN	C13H14
2-methylnaphthalene	2-MN	C11H10	1,2,7 trimethylnaphthalene	TMN	C13H14
1-methylnaphthalene	1-MN	C11H10	1,2,6 trimethylnaphthalene	TMN	C13H14

2-ethylnaphthalene	2-EN	$C_{12}H_{12}$	1,2,4-trimethylnaphthalene	TMN	$C_{13}H_{14}$
1-ethylnaphthalene	1-EN	$C_{12}H_{12}$	1,2,5 trimethylnaphthalene	TMN	$C_{13}H_{14}$
2,6 + 2,7 dimethyl naphthalene	DMN	$C_{12}H_{12}$	fenanthrene	Ph	$C_{14}H_{10}$
1,3 + 1,7 dimethylnaphthalene	DMN	$C_{12}H_{12}$	3-methylphenanthrene	3-MP	$C_{15}H_{12}$
1,6 dimethylnaphthalene	DMN	$C_{12}H_{12}$	2-methylphenanthrene	2-MP	$C_{15}H_{12}$
1,4 + 2,3 dimethylnaphthalene	DMN	$C_{12}H_{12}$	9-methylphenanthrene	9-MP	$C_{15}H_{12}$
1,5 dimethylnaphthalene	DMN	$C_{12}H_{12}$	1-methylphenanthrene	1-MP	$C_{15}H_{12}$
1,2-dimethylnaphthalene	DMN	$C_{12}H_{12}$	dibenzothiophene	DBT	$C_{12}H_8S$
1,3,7 trimethylnaphthalene	TMN	$C_{13}H_{14}$	4-methyldibenzothiophene	4-MeDBT	$C_{13}H_{10}S$
1,3,6 trimethylnaphthalene	TMN	$C_{13}H_{14}$	2 + 3- methyldibenzothiophene	2+3-MeDBT	$C_{13}H_{10}S$
1,3,5-trimethylnaphthalene	TMN	$C_{13}H_{14}$	1-methyldibenzothiophene	1-MeDBT	$C_{13}H_{10}S$

1.2 Biomarkers: site characterisation.

The chemistry of petroleum is inherently associated with the organic matter from which it is derived. This inheritance is mainly reflected in biomarkers, compounds that can be directly linked to their biological precursors and whose skeleton is preserved in such a way that it is recognisable despite being subjected to high pressures and temperatures (Peters et al. 2005; Costa de Sousa et al. 2022). Compared to n-alkanes and PAHs, the concentrations of biomarkers are low, usually in the order of parts per million. However, they can be detected by using gas chromatography coupled mass spectrometry (GC/MS; Walters et al. 2018; Rodriguez et al. 2020). Due to the variety of geological conditions under which oil forms, each type of crude oil exhibits a unique 'biomarker' or 'protobiogenic' fingerprint that identifies it. That is, a unique combination of molecules that varies to a greater or lesser extent depending on the depositional conditions that governed the basin, reservoir and/or oil well so that they can be individualised (Adedosu et al. 2012; Lorenzo et al. 2018). In addition, this characteristic pattern of biomarkers allows suggesting the nature of organic matter, depositional environment, thermal maturity, lithology of the source rock (Fang et al. 2019). This research proposes that a preliminary database of the respective oil wells could be generated from the determination of biomarkers in 15 crude oil samples obtained from various reservoirs in the Austral Basin. This would provide information on the type of source rock and organic matter, the presence or absence of oxygen, and whether paleobiodegradation processes occurred, among others. The characterisation of oil reservoirs through unconventional geochemical studies using biomarkers has been little developed in the Austral Basin (Cagnolatti and Curia 1990; Cagnolatti et al. 1996). This is partly because many oils extracted from its reservoir units are of high to very high thermal maturity and generally have low biomarker concentrations (Rodriguez et al. 2008). Therefore, a deeper understanding of the productive systems through these analyses whenever possible would allow the development of tools to improve reservoir production in this basin, i.e., a complementary contribution to the results generated in other disciplines of this science such as Geophysics, Inorganic Geochemistry and Organic Geochemistry that derive in a comprehensive study of the analysed reservoir.

1.2.1 Diagnostic relationships.

The absolute concentration of biomarkers, n-alkanes and PAHs is determined in crude oil samples when working with reference standards (Peters et al. 2005; Lundberg 2019). However, due to the costs involved, relative compositions can also be determined. The complex nature of oil means that concentration values oscillate between replicates of the same sample, therefore, to overcome this drawback, parameters known as RDs are calculated

(Table 7; Wang et al. 2006; Yang et al. 2023). With this procedure, although the independent concentrations of the analysed molecules fluctuate, the RDs are assumed to remain constant. This allows samples from the same well to be compared for variability over time or crudes from different wells and/or reservoirs to be characterised geochemically (Killops and Killops 2005). The use of RDs complements existing crude characterisation methods, but has its own advantages (Lundberg 2019). The distribution of n-alkanes, aromatics and biomarkers is source-specific, i.e. they differ from crude oil to crude oil. Furthermore, the determined DRs are relative proportions and are subject to little interference from fluctuations in the concentration of individual compounds. Therefore, they can more accurately reflect the differences of the analysed compounds between samples (Kienhuis et al. 2019). Table 7 presents some of the RDs used in this study to characterise each of the crude oils and assess their stability over time from the wells from which they were extracted.

Table 7. Diagnostic ratios used in the biomarker study.

RDs	Formula	Utility
P/F	= pristane / fitane	depositional environment
$P/n\text{-}C17$	= pristane / heptadecane	biodegradation
$F/n\text{-}C18$	= phytane / octadecane	biodegradation
$n\text{-}C29/n\text{-}C17$	= nonacosane / heptadecane	type of organic matter
$H29 / H30$	= norhopane $C29$ / hopane $C30$	type of lithology
$M30 / H30$	= moretan / hopane $C30$	type of lithology
IMP	= (2-MP + 3-MP) x 1,5 / P + 9-MP + 1-MP	state of maturity
Rc	= 0,55 x IMP + 0,44	state of maturity

IMP = methylphenanthrene index, MP = methylphenanthrene, Rc = calculated vitrinite reflectance.

1.3 Biomarkers: environmental stability.

Oil is the most widely used non-renewable natural resource by humans. Many industrial activities use it as a source of energy or raw material for thousands of everyday or industrial substances and products (Cortes et al. 2010). Fossil fuel deposits are not evenly distributed around the globe, but demand for fossil fuels is high, especially in most industrialised countries, and international transport around the world requires mainly ships. Consequently, this excessive consumption of oil generates negative impacts on humans and the environment, either indirectly, through CO_2 emissions, or directly through possible spills of crude oil and/or its derivatives (Zhang et al. 2015). Part of the current environmental legislation seeks to establish the degree of contamination of an ecosystem subjected to industrial activity and to determine its origin (Han et al. 2018). In the particular case of Argentina, this situation is contemplated in the Hydrocarbons Law (17.319) for oil pollution. Due to this, the implementation of geochemical characterisation of hydrocarbons from different sources would serve to respond to environmental problems linked to the accidental and/or intentional spillage of this resource. In addition, it would include those illicit acts involving the theft of oil at any point of the facilities belonging to the operators that during the process could lead to the first point (Orta-Martínez et al. 2018). This would allow the oil companies involved in the incidents to be held responsible for the damage caused by remediating the affected sites and paying the respective fines. On the other hand, anyone suspected of the theft, possession, and illegal distribution of crude oil could be prosecuted (Arekhi et al. 2021). In other words, this can be extended to the fraudulent marketing of crude oil and derivatives because it allows them to be identified under conditions of dubious provenance and/or seizure (Yavari et al. 2015). Finally, it would generate a better understanding of the behaviour of hydrocarbons in ecosystems and better monitoring of weathering processes in a wide variety of environmental conditions (Rodríguez et al. 2018).

1.3.1 Meteorisation.

When an oil spill occurs, it initiates a natural defence mechanism in affected ecosystems known as weathering, which promotes the natural remediation of areas affected by oil components (Wang et al. 2007). Weathering is a physical, geochemical and biological process that alters the composition of crude oil (Salat et al. 2020). In both petroleum geochemistry and environmental impact studies, knowledge of mechanisms, residence times and chemical intermediates are essential to describe the scientific and technical basis of this phenomenon (Reyes et al. 2014). Recent studies report on the mechanisms involved in oil degradation, and it is well known that each hydrocarbon has unique steps in its degradation process. These processes are not only associated with the physicochemical characteristics of the crude oil, but also influenced by biophysico-geochemical factors around the affected area (Ron and Rosenberg 2014). All oil components, including biomarkers, can be physically and geochemically altered by biotic and abiotic factors even in the reservoir rock, from exploratory stages to transport and possible spills (Joo et al. 2013). This weathering process includes phenomena such as biodegradation, dispersion, emulsification, evaporation, photooxidation and deposition that act synergistically, breaking down, deteriorating or oxidising organic molecules. Adsorption of oil or its derivatives is also possible when added to particulate matter suspended in the water column or distributed in the soil (Cai et al. 2013; Garcia et al. 2019). A brief description of the processes mentioned when crude oils are spilled over water bodies or soil-type systems is presented in Table 8.

Table 8. Processes acting on oil spills. Adapted from Reyes et al. 2014.

Process	Type	Mechanism	Effect
Adsorption	Physicochemist	Electrostatic interactions	Adhesion of crude oil to suspended particles.
Biodegradation	Biochemist	Oxidation/decomposition by micro-organisms	Physico-chemical modifications of molecules.
Dispersion	Physicist	Water turbulence	Separation of the oil slick into droplets.
Dissolution	Physicochemist	Water temperature and turbulence	Solubilisation of oil components.
Emulsification	Physicochemist	Water temperature and turbulence	Drops of crude oil in water (metastable mixtures).
Evaporation	Physicochemist	Water temperature and turbulence	Volatilisation of crude oil components.
Photooxidation	Physicochemist	Action of light, especially ultraviolet radiation and oxygen	Break-up of hydrocarbons into small molecules.

Oil weathering can be assessed in three different ways. Firstly, by the presence or absence of its components. Secondly, by the relative molecular abundance of the oil. Finally, parametrically, using relationships between heights, areas or concentrations of biomarkers and/or compounds of geochemical interest known as RDs (Stout and Wang 2016). This forms the basis of environmental analyses related to oil spills, as biomarkers have the ability to withstand weathering. In addition, they can link potential sources of contamination or differentiate between different types of crude oil involved in these events (John et al. 2018). In most cases, the coincidence or non-coincidence of biomarker distribution is strong evidence for positive or negative correlation between spilled crude oil and suspected sources. However, the above is not conclusive, because in areas of weathering, distribution patterns are modified. This results in a loss of original information, making correlation more difficult and necessitating the use of other tools (Wang and Fingas 2003; Lobao et al. 2022). Weathering affects the status of different biomarkers and compounds in the crude oil, and their molecular abundance indicates the deterioration and residence time of these compounds (Stout and Wang 2007).

In general, biomarkers retain a great deal of information from the original oil, and this structural similarity reveals very specific information about the source of the spilled oil. Sometimes the crude samples being compared are chemically similar, i.e. the hydrocarbon profiles exhibit little variation between them, which also presents a difficulty in solving the problem. In addition, there are situations where there is more than one suspected source responsible for an accidental or intentional oil spill (Wang et al. 2000; Oliveira et al. 2020). On the other hand, the chemical characteristics of the source and weathering persistent biomarkers are a critical input for environmental forensic investigations when determining the source and correlating crude oil (Wang et al. 2001; Yang et al. 2023). These issues can be resolved by examining and comparing biomarker patterns and profiles because they are widely used for oil correlation in environmental forensic studies.

1.4 Geological framework.

The Austral Basin formed during the Triassic-Jurassic and is located at the southern end of Patagonia (Fig. 4). To the north it is bounded by the Deseado Massif, a structural high in north-central Santa Cruz Province. To the northwest it connects with the western part of the Golfo San Jorge basin which lies between northern Santa Cruz province and southern Chubut province (Ramos et al. 2019). A quarter of the basin is in Chilean territory, on the island of Tierra del Fuego and north of the Strait of Magellan, and the rest in Argentina covering an area of 162,000 km^2 (Aramendia et al. 2018). It has a stable shelf (continental territory) that covers a strip approximately 600 km long by a maximum of 150 km wide, attached to the maritime coastline of the provinces of Santa Cruz and Tierra del Fuego (Barberón et al. 2015). It is followed by the offshore sector, which covers the entire maritime coastline, from the coast to the Dungeness High in Argentina and part of the Strait of Magellan in Chile. It continues with an oceanic slope and basin located in the central west of the province, documenting abundant hydrocarbon manifestations in the deepest sector of the basin (Zerfass et al. 2017).

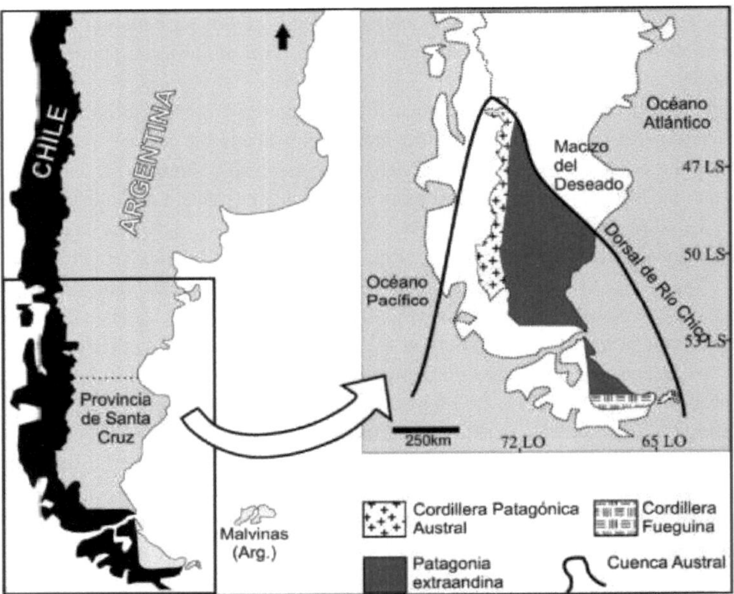

Figure 4. Geographical location of the Austral Basin and the geological provinces in which it develops. Taken from Richiano et al. 2012.

The development of the Austral Basin is strongly linked to the geodynamic evolution of the curvature of the southernmost Andes (concave side), while the hinge sector contains the thickest depocentre (Diraison et al. 2000). After late Palaeozoic-Triassic cratonic accretion and basement consolidation (Giacosa et al. 2012), a Jurassic extensional event established the basin substrate. During Gondwana extension, the Green Rocks oceanic basin opened along the continent margin (Calderón et al., 2016). In addition, a regional N-S to NE-SW oriented rift system with synrift infilling of continental and volcanic sequences affected the entire basin (Cuitiño et al. 2019). The rift stage of sedimentation in the Austral Basin occurred during the late Upper Jurassic and early Cretaceous. This extensional phase corresponds to the Tobifera Fm, which has a variable thickness ranging from 100 to more than 1000 m, and is mainly composed of volcano-sedimentary sequences (Poiré and Franzese, 2010). Overlying this complex is the Springhill Fm, which represents the initial transgressive infill of the pits and semifossae (Cuitiño et al. 2019). This unit was first defined in subsurface studies and represents the most important conventional reservoir in the Austral Basin (Schwarz et al. 2011). After the rifting phase in the Late Jurassic, a transgressive tectonic cycle characterised by the deposition of the Lower Inoceramus Fm composed of organic-rich shales began in the basin (Gallardo 2014). The subsidence phase continued during the Middle Cretaceous with the deposit on of the Green Marl Fm, represented by calcareous shales and marine argillites, signalling the beginning of a new tectonic cycle. The Tertiary reservoir units of the basin are represented by typical sandstones in two units: Lower Magallanes (Paleocene-Eocene) and Upper Magallanes (late Eocene - early Miocene). The former rests unconformably on the Cretaceous and is composed of glauconitic and quartz sandstones and shales, deposited in a marine environment (Mpodozis et al. 2011). The second is composed of conglomerates, sandstones, shales and lignite beds, deposited in a fluvial to marginal marine environment

14

(Barberón et al. 2015). The overlying Miocene formation is represented by continental deposits of the Santa Cruz Fm. Figure 5 shows a stratigraphic diagram of the Austral Basin for different localities in the basin, showing the traditional subsurface units.

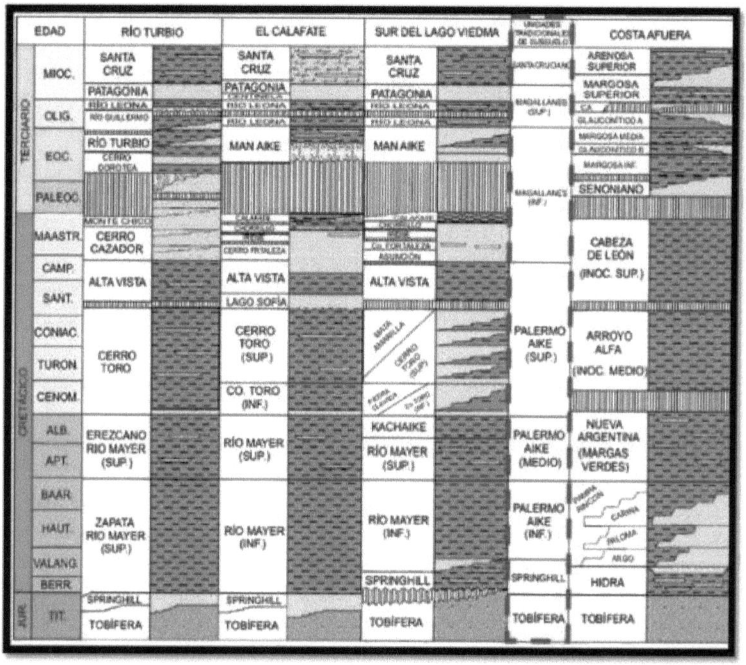

Figure 5. Stratigraphic diagram of different localities in the Austral Basin. Adapted from Schiuma et al. 2018.

Based on some evidence from source rock-hydrocarbon correlations, six petroleum systems have been recognised in the Austral Basin, of which three are classified as known: "Palermo Aike/Inoceramus Inferior - Springhill", "Margas Verdes - Magallanes Inferior" and Tobifera - Tobifera/Springhill (Rodríguez et al. 2008). The Springhill Fm is the main reservoir rock of the basin, known since the 1940s to have provided virtually all of the hydrocarbon resources discovered to date, however, there are several oil and gas reservoir units in the Lower Magallanes Fm (Cagnolatti and Miller 2002). Recent geochemical studies of petroleum from different reservoirs in the Springhill Fm. and Lower Magallanes Fm. show that they have similarities and differences. In several cases it is difficult to determine whether these differences are due to the contribution of different source rocks or to other factors such as: variations in source facies, thermal maturity, alteration processes in the reservoir, minor changes and contributions during migration and finally sampling problems (Rodriguez et al. 2008).

During the evolution of the basin there was a progressive burial of the source rocks that favoured the generation of hydrocarbons. The Lower Palermo Aike/Inoceramus Inferior Fm was characterised by a mixed kerogen (type II/III) which was determined from the results of hydrogen and oxygen indices obtained by rock pyrolysis - eval (Belotti et al. 2014). The accumulation of this type of organic matter is related to the anoxic impact on the subsidence belt of the South American slab (Legarreta & Villar 2011). In this sense, the source pelites of the Margas Verdes Fm have been correlated with marine petroleum in the southern sector of the basin (Pittion and Gouadain, 1992; Pittion and Arbe 1999). On the other hand, the source rock known as the Tobifera Series has associated facies that accumulated under restricted lacustrine conditions giving rise to Type III and to a lesser extent Type I kerogens (Legarreta & Villar 2011).

1.5 Problem statement.

The gas-bearing Austral Basin has numerous light hydrocarbon deposits as a result of its complex geological history. It is currently undergoing ongoing development plans to discover and produce its oil reserves. However, information related to crude oil biomarkers has been little developed in the last 30 years in this area. This paper hypothesises that each crude oil extracted from an oil well has a unique and stable chemical signature formed by a set or pattern of biomarkers that will allow geochemical characterisation of the crude oil samples. Furthermore, assessing their behaviour over time will or will not guarantee their reproducibility. Finally, a cluster analysis will be developed to determine the degree of relatedness of the crude oils across the basin. If these results are positive, they have the potential to generate a database that will allow the traceability of each of the crudes in the basin, with its consequent application in litigation for the theft, transport, storage and distribution of stolen hydrocarbons. On the other hand, the recalcitrance of these biomarkers to weathering will also be evaluated when the crudes are exposed to laboratory conditions for a year in reactors with seawater and soil. If these molecules remain stable for the defined time, their behaviour could be extrapolated to situations of environmental contamination by accidental or intentional spills of crude oil, thus finding those responsible who should be brought to justice.

1.6 General objective.

To evaluate the distribution of biomarkers in oil from different oilfields in the Austral Basin, Santa Cruz province, in order to generate information that can be applied to identify possible spills in the environment.

1.6.1 Specific objectives.

SO1: Evaluate biomarker profiles in crude oil samples from the Del Mosquito, Campo Indio, Cañadón Salto and La Maggie reservoirs, linked to the Springhill Fm, and the Agua Fresca and María Inés reservoirs associated with the Lower Magallanes Fm.

SO2: Evaluate the origin of the crude oils associated with the different reservoirs studied in terms of the biomarker profiles found.

SO3: To know the environmental stability of the different biomarkers found in the crude oil samples studied.

1.7 Hypothesis.

H1: The set of biomarkers in a hydrocarbon sample constitutes a unique fingerprint or chemical signature for each crude oil and its composition does not vary within a geological reservoir on human timescales.

H2: Biomarkers have the capacity to resist weathering processes to a lesser or greater extent depending on the environmental conditions, type of system impacted and time of permanence of the crude oil in the system.

H3: Biomarkers have the potential as a tool to solve problems linked to accidental and/or intentional oil spills and, in addition, to solve oil theft disputes.

2 MATERIALS AND METHODS

This chapter describes the procedures that were carried out in order to achieve the objectives set out in this research.

2.1 Crude oil samples.

For the development of this study, 15 oil wells were selected from six strategically located fields in the Austral Basin (Fig. 6). Almost all of the crude oil samples were supplied by Compañía General de Combustibles (CGC). Only those hydrocarbon samples extracted from the Del Mosquito field were provided by Fomicruz Sociedad del Estado. In addition, only four crude oil samples were obtained from five wells in the period of one year, since due to unforeseen situations for the others, only 2 or 3 samples were accessed. Table 9 shows in detail the name and number of samples, their respective reservoirs, geological formations and drilling depths.

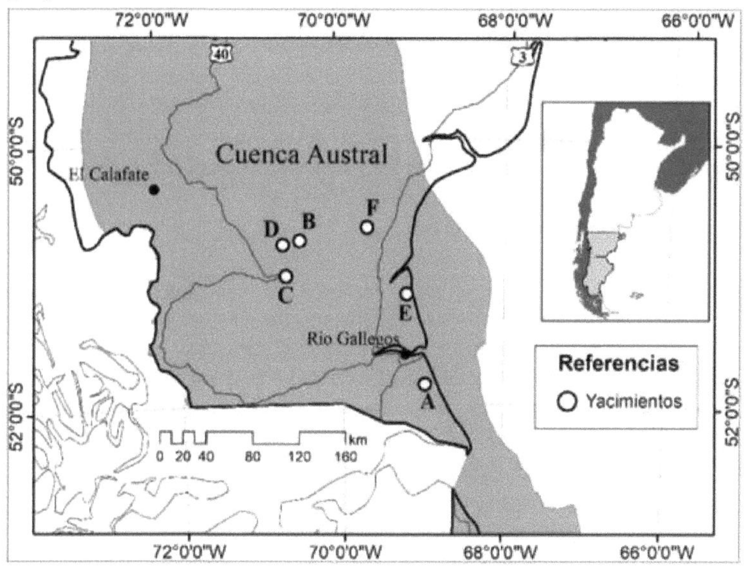

Figure 6. Extract from the map of the Austral Basin showing the studied deposits.
Del Mosquito (A), Agua Fresca (B), María Inés (C), Campo Indio (D), Cañadón Salto (E) and La Maggie (F).

Table 9. Description of the samples under study.

Name	Samples	Site	Training	Punching Coordinates	Coordinates
AC	4	From the Mosquito	Springhill	1300 m	51°51' 16" S	68° 57' 42" W
AS	4	From the Mosquito	Springhill	1300 m	51° 53' 03" S	68° 58' 31" W
AN	4	From the Mosquito	Springhill	1300 m	51° 47' 47" S	68° 59' 01" W
BI	3	Fresh Water	Lower Magellan	1400 m	50° 44' 41" S	70° 30' 60" W
BM	2	Fresh Water	Lower Magellan	1400 m	50° 46' 01" S	70° 30' 20" W
BS	4	Fresh Water	Lower Magellan	1400 m	50° 45' 29" S	70° 33' 11" W
CI	2	María Inés	Lower Magellan	1600 m	51° 02'21 "S	70° 37' 38" W
CO	3	María Inés	Lower Magellan	1650 m	50° 59' 58" S	70° 45' 18" W
DB	3	Campo Indio	Springhill	3000 m	50° 47' 01" S	70° 43' 18" W
DA	4	Campo Indio	Springhill	3000 m	50° 46' 43" S	70° 43' 34" W

EB	2	Cañadón Salto	Springhill	1320 m	51° 10' 34" S	69° 11' 24 "W
EA	2	Cañadón Salto	Springhill	1320 m	51° 08'54 "S	69° 12' 44" W
FI	3	The Maggie	Springhill	1500 m	50° 38' 47" S	69° 40' 53" W
FM	2	The Maggie	Springhill	1500 m	50° 39' 18" S	69°41' 23 "W
FS	2	The Maggie	Springhill	1500 m	50° 39' 36" S	69°41' 11 "W

The samples were collected and transported to the laboratory in previously washed one-litre glass bottles (amber colour). The caramel colouring of the bottles prevented the development of photochemical reactions on the hydrocarbons. During the filling of the bottles, the formation of an air chamber was avoided to minimise the impact of oxygen on the stability of the samples and the occurrence of biodegradation processes stimulated by an oxic microenvironment. Finally, they were stored in a dark and dry place at room temperature until the analyses were performed within 72 hours.

The techniques used to condition and analyse the crude oil samples were divided according to the specific research objectives because they involved different working strategies. Firstly, experimental techniques are presented to determine the biomarker profiles of the crude oils studied and establish their degree of relatedness according to the reservoir, the geological formation and the number of samples taken during the experiment. Secondly, the methodologies necessary to evaluate the environmental stability of the biomarkers of one of the crude oil samples in soil and seawater are developed.

2.1.1 Determination of compounds of interest.

2.1.1.1 Conditioning of crude oil samples.

Taking into consideration the Texas TNRCC standard (1006), the samples were conditioned by two processes according to the type of chromatographic analysis used. The first involved dilution of the crude samples (1/100) with n-pentane and subsequent cleaning with one gram of silica gel to obtain a translucent solution. From it, the total ion chromatogram (TIC) was subsequently obtained, and therefore defined as TIC extract. Secondly, the crude oils were fractionated by solid-liquid column adsorption chromatography into saturated hydrocarbons, aromatic compounds and resin-asphaltenes. The samples were subjected to separation on a glass column (20 cm x 1.2 cm) packed with 3 g of silica gel (activated at 150 °C for a period of 24 hours in an oven), to which 50 mg of activated sodium sulphate and 50 mg of activated alumina were added on top. Approximately 100 µL of crude oil was seeded on the column, then successively eluted with 10 mL of n-pentane and 10 mL of dichloromethane, to obtain the aliphatic and aromatic extracts, respectively, with the heavy fractions known as resins and asphaltenes being retained on the column (Fig. 7). The aliphatic and aromatic fractions were separately concentrated to 0.5 mL under nitrogen stream, transferred to a chromatography vial, and stored at -15 °C until analysis.

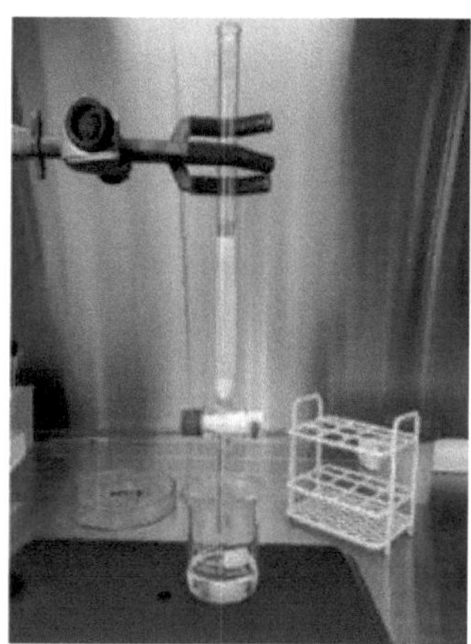

Figure 7. Column chromatography of one of the crude samples.

2.1.1.2 Chromatographic analysis of crude oil samples.

Based on the publications of Stashenko et al. (2014) and Tomas et al. (2023), for each crude sample, 1 µL of the TIC extract in split mode and 1 µL of the aliphatic and aromatic extracts in splitless mode were injected into a gas chromatograph (Fig. 8). The separation was performed on an Agilent model 7890A, with an Agilent mass spectrometry detector (model 5975C). A 30 m long HP5ms column with an internal diameter of 0.32 mm and a film thickness of 0.25 µm was used. The injector temperature was set to 290 °C and helium was used as carrier gas with a flow rate of 1.2 mL.min^{-1} . The temperature programme used was as follows: initial temperature of 55 °C for 2 min, followed by a ramp of 6 °C.min^{-1} until 270 °C was reached, passing directly to another ramp of 3 °C.min^{-1} until 300 °C was reached, which was maintained for 17 min. The total run time was 65 min. The mass detector was used with an ion source and transfer line temperature of 230 °C and 180 °C respectively, and an impact energy of 70 eV.

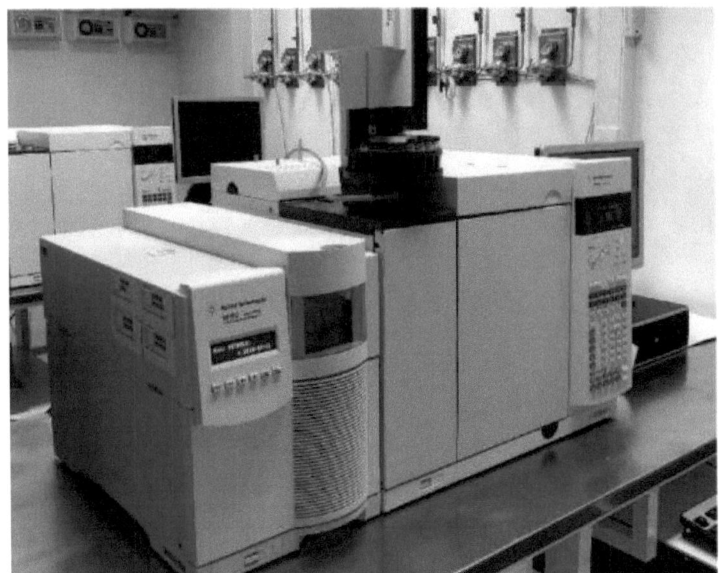

Figure 8. GC/MS of the FRSC - UTN.

Mass scanning between 30 and 400 amu was performed to obtain the total ion chromatogram (TIC) in order to determine the acyclic isoprenoids pristane and phytane and the distribution of n-alkanes from the TIC extract. The Full Scan (all ion scan) and SIM Scan (selected ion monitoring) modes were used to analyse the ions m/z = 128 (naphthalene), 142 (methylnaphthalenes), 156 (dimethylnaphthalenes), 170 (trimethylnaphthalenes), 178 (phenanthrene), 170 (trimethylnaphthalenes), 178 (phenanthrene), 170 (trimethylnaphthalenes) and 178 (phenanthrene), 178 (phenanthrene), 184 (dibenzothiophene and tetramethylnaphthalenes), 192 (methylphenanthrenes) and 198 (methyldibenzothiophenes) in the aromatic fraction, and the ions m/z = 191 (terpanes) and 217 (steranes) in the aliphatic fraction, respectively. In this way, the fragmentograms of interest corresponding to each sample were obtained using the software "MSD ChemStation Data Analysis Application". From the retention times, reference literature and manual integration of the peaks, the n-alkanes, PAHs and biomarkers described in sections 1.1.2 and 1.1.3 were identified.

2.1.1.3 Data normalisation.

The chemical species of interest were integrated in the corresponding TICs and fragmentograms considering the area under the curve of the peaks that identify them. The values obtained were summed to generate a total that equals 100 % of the compounds of interest. Then, each value was divided by the total to obtain the relative abundance (RA) of the n-alkanes, pristane and phytane in the TIC and of the terpanes, steranes and PAHs in their corresponding fragmentograms (Equation 1). The calculations were performed in a Microsoft Office Excel spreadsheet designed for easier manipulation of the data. Therefore, this spreadsheet was validated and its use was complied with.

$$(1) \quad AR = \left(\frac{\text{área bajo la curva del compuesto químico de interés}}{\text{sumatoria de las áreas de los compuestos químicos de interés}} \right) . 100$$

area under the curve of the chemical compound of interest
sum of the areas of the chemical compounds of interest

2.1.1.4 Sampling variability of the same well over time.

Once the RAs of the chemical species of interest for n-alkanes, terpanes, steranes and PAHs were determined for the 15 crude oils studied over the four samplings, the RDs were calculated as shown in Table 7. To verify the reproducibility of the results obtained, the relative standard deviation (RSD) of the RAs and RDs was calculated (Equation 2). If the RSD remains below 14 % in all cases as established in the European standard CEN/TR 15522-2, it can be established that there is no variability between samples from the same oil well and their values can be averaged to work with greater robustness. It should be noted that this percentage is justified due to working with samples that are in contact with various environmental compartments, such as sediments, groundwater, formation water, storage tanks, among others (Kienhuis et al. 2019).

$$(2) \quad RSD = \left(\frac{\text{desviación estándar de la AR o RD}}{\text{promedio de la AR o RD}} \right) . 100$$

standard deviation of the RA or RD
average RA or RD

2.1.1.5 Multivariate statistics.

To establish the degree of clustering of crude samples from the same well over time, multivariate statistics were used. Using Past 3 software, a principal component cluster analysis (PCA) was carried out on the results obtained. To assemble the data matrix, 16 diagnostic indices were considered, some of which are shown in Table 7. These parameters were selected from common biomarkers in the samples that were part of this thesis.

2.2 Geochemical characterisation.

From the data obtained and considering certain DRs and/or RAs, a series of graphs were generated and are detailed below:

- $P/n\text{-}C_{17}$ vs P/F (Peters et al. 2005).
- P/n-C17 vs F/n-C18 (Shanmugam 1985).
- Relative percentages of 4-MeDBT / 2 + 3-MeDBT / 1-MeDBT as a function of each sample (Killops and Killops 2005).
- DBT/Ph vs P/F (Hughes et al. 1995).
- Ternary diagram of steranes (Moldowan et al. 1985).

These plots are a valuable tool for the geochemical characterisation of crude oils, i.e. they allow determining the type of kerogen in the crude oil and the level of palaeobidegradation if any (López and Lo Mónaco 2017; Chen et al. 2018). In addition, they give an idea of the nature of the source rock in which the hydrocarbons originated and what the oxygen concentration was like when the precursor organic matter settled (López et al. 2019).

2.3 Environmental stability.

For the environmental stability analysis under laboratory conditions, the sample extracted from the AS oil well was selected, which had an API gravity of 32° at 15 °C (light crude) and a salinity of 52 g/m^3 . This decision was made because it was one of the samples with the highest amount of terpanes and steranes, and also had high AR values. On the other hand, a Patagonian soil sample of approximately 5 kg was taken in the vicinity of the city of Rio Gallegos, Santa Cruz, Argentina, at a depth of between 0 and 15 cm (A horizon) with respect to the surface, after removing the leaf litter present. The sample was sieved with a 2 mm pore mesh and stored refrigerated until the time of study. The soil was characterised by a sandy loam texture, and the following physicochemical properties: pH 7.6; chloride 37.4 ppm; sulphate 48.7 ppm; bicarbonate 45.7 ppm; carbonate < 1 ppm; calcium 40.8 ppm; magnesium 12.2 ppm; nitrite 0.52 ppm; nitrate 18.1 ppm; ammonium 0.05 ppm; phosphate < 1 ppm; conductivity 84.5 μS.cm^{-1} ; moisture 1.07 %; organic matter 0.98 %; inorganic matter 99.02 %; bulk density 1.31 g.cm^{-3} ; real density 2.46 g.cm^{-3} ; porosity 46 %; total hydrocarbons < 1 mg.kg^{-1} ; total aerobic bacteria (TAB) 5.30 x 10^5 CFU.g^{-1} ; hydrocarbon degrading bacteria (HDB) 8.90 x 10^3 CFU.g^{-1} .

2.3.1 Artificial weathering test of crude oil in seawater and soil.

The study of crude oil weathering in seawater was carried out in 30 laboratory-scale reactors (300 mL capacity clear glass flasks) monitored for a period of one year, evenly distributed over ten stages and with manual stirring once a week for a period of 1 min (Figure 9). In each of them, 100 mL of seawater from Punta Loyola (51°36'25" S, 69°01'08" W) and 1 mL of crude oil were added. On the other hand, a similar procedure was followed for the lithospheric systems, but with a major difference. In each reactor, 10 g of litter-free soil from the horizon between the surface and the first 15 cm of depth were added. One millilitre of crude oil was then carefully distributed into each reactor with a Pasteur pipette and spatula to ensure homogeneous distribution throughout the soil. In both cases the reactors were left open, i.e. without a lid, at the average ambient laboratory temperature of 20 °C so that phenomena such as photo-oxidation, evaporation and biodegradation would modify the chemical composition of the hydrocarbon. Time zero (T0) was considered for the crude oil samples incorporated into the three initial reactors and recovered immediately for analysis, i.e. zero days after the start of the experiment. The remaining stages, also comprising three reactors, are detailed below: 7, 14, 21, 28, 28, 60, 120, 180, 270 and 365 days constitute the analysis times T1, T2, T3, T4, T5, T6, T7, T8 and T9, respectively.

Figure 9. Raw and seawater systems in glass jars.

Once the time intervals for each stage were completed, the crude oil incorporated into the reactors was recovered with seawater by adding 10 mL of n-pentane to each reactor in a single extraction step. Subsequently, the organic phase was obtained using a 200 mL decanting vial, recovered and transferred to a vial. The volume contained in the vials was divided in two, one fraction was stored refrigerated and the other was subjected to the same treatment described in sections 2.2.1.1 and 2.2.1.2. With regard to the crude samples in soil, once the time of permanence in such conditions was completed, the hydrocarbons were extracted from the solid system by means of accelerated solvent extraction (ASE) with the Dionex ASE 150 "Thermo Scientific" equipment (Figure 10). The extraction conditions were: 175 °C cell temperature, 1500 psi pressure, 5 min extraction time and one cycle (EPA 3545A). The extract obtained was concentrated under nitrogen stream to a volume of 10 mL, an aliquot of one millilitre was collected in a vial and named TIC extract (Stashenko et al. 2014). The remaining volume was further concentrated under nitrogen stream to 0.5 mL, transferred to a chromatography vial, and followed the same methodology as described in 2.2.1.1 and 2.2.1.2. The ions mentioned in 2.2.2.2.1 were also added to the GC/MS analysis.

Figure 10. Dionex ASE 150 equipment.

To eliminate the effect of analytical variability, all artificially weathered crude oil samples were analysed in triplicate in both seawater and soil in each of the 10 time periods defined for the experiment. Based on the literature published by Fernández-Varela et al. (2010), 12 RDs (Table 10) derived from hopanes and steranes commonly used for the identification of sources of hydrocarbon contamination were calculated to assess their recalcitrance to weathering processes. On the other hand, based on works by Wang et al. (2000) and Lemkau et al. (2010), other diagnostic indices generated from n-alkanes, pristane, phytane and PAHs were considered (Table 10). It should be noted that RSD was used as an indicator to assess the variability of DRs over the course of the experiment (Zhang et al. 2015).

Table 10. Diagnostic relationships used in the environmental stability analysis.

Hopanos and esteranos	Alkanes, isoprenoids and PAHs	Additional information
Ts/H30	P/n-Ci7	Biodegradation
G30/H30	F/n-Ci8	Biodegradation
M30/H30	P/F	Biodegradation
Ts/ Tm	$(n\text{-}C13 + n\text{-}C14) / (n\text{-}C25 + n\text{-}C26)$	Evaporation
M30/H29	N0 + Ni / N2	Evaporation and dissolution
H31(R)/H31(S)	2-MP /1-MP	Photooxidation
D27 βα(R)/H30	4-MeDBT / I-MeDBT	Biodegradation
D27βα(S)/D27βα(R)	2 + 3-MeDBT/1-MeDBT	Biodegradation
S28αββ(R+S)/H30		
D27βα(R)/S29αααα(S)		
S29αααα(S)/H30		
S29αααα(S)/S29ααββ(R+S)		

Ts/H30 = trisnorneohopane/hopane C30, G30/H30 = gammacerane/hopane C30, M30/H30 = moretan/hopane C30, Ts/Tm = trisnorneohopane/trisnorhopane, M_{30}/H_{30} = moretan/hopane C_{29}, H_{31} (R)/H_{31} (S) = homohopane C_{31} (R)/(S), D_{27} βα(R)/H_{30} = diasterano C_{27} /hopano C , D_{3027} ßa(S)/D_{27} βα(R) = diasteranos ßa(S)/(R), S28aßß(R+S)/H_{30} = ergostanos/hopano C_{30} , D_{27} ßa(R)/S_{29} aaa(S) = diasterano/estigmastano, S_{29} aaa(S)/H_{30} = estigmastano/hopano C_{30} , S_{29} aaa(S)/S_{29} aßßß(R+S) = stigmastanes aaa(S)/aßßß(R+S), P/n-C_{17} = pristane/heptadecane, FZa-C_{18} = phytan/octadecane, P/F = pristane/phytane, (n-C +n-C_{1314})/(n-C +n-C_{2526}) = tridecane + tetradecane/pentadecane + hexadecane, N0 + N1 / N2 = naphthalene + methylnaphthalene/dimethylnaphthalene, 2-MP /1-MP = 2/1- methylphenanthrenes, 4-MeDBT / 1-MeDBT = 4/1-methyldibenzothiophenes, 2 + 3-MeDBT/1-MeDBT = 2 + 3/1-Methyldibenzothiophenes.

To demonstrate the environmental stability of the biomarkers against weathering under laboratory conditions, the values of the DRs as a function of time were analysed through the RSDs. The RSD of each DR presented in Table 10 was calculated according to equation 3, where SD is the standard deviation of the DRs between the times from T0 to T9 and Prom is the average of the DRs between the times from T0 to T9. For the RDs to be considered stable over the defined time, their RSDs must be less than 5 % (Zhang et al. 2015).

$$(3) \quad RSD = \left(\frac{SD\ (T0{:}T9)}{Prom\ (T0{:}T9)} \right) . 100$$

The chromatograms and fragmentograms obtained over the time frame established in this type of study make it possible to quickly observe whether or not modifications occur in the crude oil samples that can be associated with the weathering processes taking place in the soil and/or seawater. In addition, a histogram was made for each time defined in this research with the respective series of PAHs and their alkylated derivatives (naphthalenes, phenanthrenes, dibenzothiophenes). The objective is to observe modifications in their relative abundances as a result of the weathering conditions to which the crude oil samples were subjected both in seawater and soil at laboratory scale.

3 RESULTS

The results obtained in the framework of this research are presented below. In the first part, the chemical characterisation by biomarkers of the 15 crude oils sampled every three months over the course of a year in six strategically distributed fields in the Austral Basin is developed. The second part presents the environmental stability study in seawater and soil at laboratory scale for the AS sample during 12 months distributed in 10 stages of analysis.

2.4 Stability of the sample over time.

The RAs of the chemical compounds of interest in no case exceeded 14 % of their RSD in the four samplings carried out over one year. For practical reasons, only the results obtained for AS crude oil are presented in Table 11. This is a complex matrix that is characterised by containing the RAs of the four samples defined as AS1, AS2, AS3 and AS4, and the averages, SDs and RSDs calculated. It is interesting to note that the RSDs for n-alkanes and PAHs ranged between 1 and 13 %, for terpanes they were approximately between 3 and 13 %, and finally, steranes presented the narrowest range with values between 4 and 12 %. In other words, the results were lower than 14 % as required by the European standard CEN/TR 15522-2 (2012; Lundberg 2019) and the crude AS did not change in one year. The remaining 14 tables are presented in Annex 1. They can be summarised by saying that the RSDs linked to the RAs of each compound of interest did not exceed 14 % for the remaining crude oils analysed over the four samplings performed.

Table 11. Relative standard deviation obtained from the relative abundances of AS crude oil.

RAs	AS1	AS2	AS3	AS4	Averages	SDs	RSDs
n-C9 = nonane	0,033	0,039	0,038	0,039	0,037	0,003	8,0
n-C10 = **dean**	0,050	0,052	0,054	0,052	0,052	0,002	3,2
n-C11 = **undecane**	0,074	0,067	0,069	0,067	0,069	0,003	4,7
n-C12 = **dodecane**	0,070	0,066	0,060	0,065	0,065	0,004	6,6
n-C13 = **tridecane**	0,066	0,064	0,064	0,064	0,064	0,001	1,6
n-C14 = **tetradecane**	0,061	0,061	0,058	0,058	0,059	0,001	2,2
n-C15 = **pentadecane**	0,056	0,054	0,053	0,062	0,056	0,004	6,9
n-C16 = **hexadecane**	0,047	0,045	0,046	0,045	0,045	0,001	1,8
n-C17 = **heptadecane**	0,040	0,043	0,041	0,041	0,042	0,001	2,8
P = pristano	0,019	0,017	0,018	0,017	0,018	0,001	5,6
n-C18 = **octadecane**	0,040	0,037	0,040	0,037	0,038	0,001	3,8
F = filano	0,009	0,008	0,009	0,009	0,009	0,001	6,1
n-C19 = **nonane**	0,040	0,041	0,041	0,040	0,040	0,001	1,9
n-C20 = **eicosane**	0,043	0,043	0,042	0,041	0,042	0,001	2,0
n-C21 = **eneicosane**	0,049	0,050	0,047	0,046	0,048	0,002	3,3
n-C22 = **docosane**	0,041	0,045	0,045	0,043	0,043	0,002	4,3
n-C23 = **trichosane**	0,052	0,052	0,052	0,050	0,051	0,001	2,5
n-C24 = **tetracosane**	0,040	0,040	0,041	0,040	0,040	0,000	0,9
n-C25 = **pentacosane**	0,041	0,044	0,043	0,042	0,042	0,001	2,4
n-C26 = **hexacosane**	0,033	0,036	0,036	0,035	0,035	0,001	3,4
n-C27 = **heptacosane**	0,031	0,032	0,035	0,036	0,034	0,002	7,1
n-C28 = **octacosane**	0,025	0,026	0,024	0,028	0,026	0,002	6,3
n-C29 = **nonacosane**	0,022	0,023	0,023	0,024	0,023	0,001	4,5
n-C30 = **triacontane**	0,018	0,016	0,020	0,019	0,018	0,002	9,2
T19 = **terpane**	0,019	0,021	0,017	0,017	0,018	0,002	9,7
T20 = **terpane**	0,021	0,022	0,021	0,019	0,021	0,001	5,4
T21 = **terpane**	0,021	0,024	0,024	0,020	0,022	0,002	9,1
T23 = **terpane**	0,022	0,024	0,0242	0,020	0,023	0,002	8,0
T24 = **terpane**	0,0153	0,017	0,018	0,017	0,017	0,001	5,9
T25 = **terpane**	0,004	0,004	0,005	0,004	0,004	0,000	6,6

T26 (R) = terpane	0,039	0,042	0,042	0,039	0,040	0,001	3,1
T26 (S) = terpane	0,003	0,004	0,004	0,003	0,003	0,000	11,3
Ts = trisnorneohopane	0,046	0,044	0,038	0,049	0,044	0,005	11,1
Tm = trisnorhopane	0,078	0,078	0,075	0,062	0,073	0,008	10,5
H29 = hopane	0,260	0,254	0,238	0,249	0,250	0,009	3,6
H30 = hopane	0,298	0,296	0,311	0,309	0,303	0,008	2,6
M30 = Moretan	0,035	0,032	0,037	0,035	0,035	0,002	5,7
H31 (S) = homohopane	0,053	0,053	0,054	0,062	0,056	0,004	7,2
H31 (R) = homo-hopane	0,028	0,029	0,029	0,032	0,030	0,002	6,1
G30 = gammaceran	0,008	0,008	0,009	0,010	0,009	0,001	12,1
H32 (S) = homohopane	0,020	0,022	0,025	0,021	0,022	0,002	9,7
H32 (R) = homo-hopane	0,015	0,016	0,018	0,020	0,017	0,002	11,8
H33 (S) = homohopane	0,009	0,008	0,008	0,007	0,008	0,001	7,9
H33 (R) = homohopane	0,004	0,003	0,004	0,003	0,003	0,000	9,4
S20 = pregnane	0,047	0,052	0,054	0,043	0,049	0,005	9,8
S21 = homopregnane	0,049	0,053	0,049	0,050	0,050	0,002	4,1
S22 = homopregnane	0,039	0,041	0,037	0,043	0,040	0,003	6,6
D27 (βs) = diasterane	0,099	0,104	0,092	0,091	0,097	0,006	6,3
D27 (βr) = diasterano	0,043	0,045	0,040	0,041	0,042	0,002	4,4
D27 (as) = diasterano	0,012	0,011	0,013	0,013	0,012	0,001	8,6
D27 (αr) = diasterane	0,012	0,011	0,014	0,013	0,012	0,001	10,1
S27 (as) = cholestane	0,041	0,039	0,046	0,047	0,043	0,004	9,9
S27 (βr) = cholestane	0,084	0,070	0,082	0,095	0,083	0,010	12,4
S27 (βs) = cholestane	0,040	0,040	0,050	0,048	0,044	0,005	11,4
S27 (αr) = cholestane	0,046	0,056	0,062	0,057	0,055	0,007	12,6
S28 (as) = ergostane	0,001	0,001	0,001	0,001	0,001	0,000	11,7
S28 (βr) = ergostane	0,019	0,017	0,021	0,020	0,019	0,002	9,1
S28 (βs) = ergostane	0,021	0,019	0,024	0,021	0,021	0,002	9,2
S28 (αr) = ergostane	0,050	0,038	0,047	0,042	0,044	0,005	12,0
S29 (as) = stigmastane	0,173	0,175	0,155	0,144	0,162	0,015	9,1
S29 (βr) = stigmastane	0,060	0,059	0,048	0,054	0,055	0,005	9,7
S29 (βs) = stigmastane	0,015	0,018	0,017	0,018	0,017	0,001	7,5
S29 (αr) = stigmastane	0,150	0,153	0,150	0,158	0,153	0,004	2,4
N = naphthalene	0,014	0,012	0,015	0,013	0,013	0,001	8,0
2 - MN = methylnaphthalene	0,057	0,059	0,066	0,052	0,059	0,006	10,2
1 - MN = methylnaphthalene	0,039	0,039	0,048	0,042	0,042	0,004	10,1
2 - EN = ethylnaphthalene	0,010	0,008	0,009	0,010	0,009	0,001	9,2
1 - EN = ethylnaphthalene	0,008	0,007	0,009	0,007	0,008	0,001	10,9
2,6+2,7-DMN = dimethylnaphthalene	0,062	0,061	0,072	0,059	0,063	0,006	9,5
1,3+1,7-DMN = dimethylnaphthalene	0,067	0,066	0,068	0,076	0,069	0,005	6,8
1,6 - DMN = dimethylnaphthalene	0,039	0,038	0,045	0,041	0,041	0,003	7,2
1,4+2,3-DMN = dimethylnaphthalene	0,035	0,041	0,042	0,040	0,040	0,003	7,5
1,5 - DMN = dimethylnaphthalene	0,011	0,011	0,013	0,012	0,012	0,001	8,8
1,2 - DMN = dimethylnaphthalene	0,022	0,024	0,029	0,027	0,026	0,003	11,3
1,3,7 - TMN = trimethylnaphthalene	0,049	0,049	0,059	0,053	0,053	0,005	9,2
1,3,6 - TMN = trimethylnaphthalene	0,045	0,046	0,055	0,052	0,050	0,005	9,8
1,3,5 - TMN = trimethylnaphthalene	0,031	0,030	0,039	0,036	0,034	0,004	12,1
2,3,6 - TMN = trimethylnaphthalene	0,023	0,019	0,024	0,021	0,022	0,002	10,4
1,2,7 - TMN = trimethylnaphthalene	0,026	0,029	0,025	0,028	0,027	0,002	6,8
1,2,6 - TMN = trimethylnaphthalene	0,005	0,005	0,005	0,005	0,005	0,000	9,4
1,2,4 - TMN = trimethylnaphthalene	0,009	0,010	0,010	0,010	0,010	0,001	6,1
1,2,5 - TMN = trimethylnaphthalene	0,079	0,073	0,067	0,066	0,071	0,006	8,7
Ph = phenanthrene	0,102	0,094	0,076	0,086	0,089	0,011	12,0
3 - MP = methylphenanthrene	0,049	0,058	0,045	0,055	0,051	0,006	11,6
2 - MP = methylphenanthrene	0,065	0,070	0,054	0,066	0,064	0,007	11,2

9 - MP = methylphenanthrene	0,060	0,067	0,052	0,062	0,060	0,006	10,4
1 - MP = methylphenanthrene	0,049	0,046	0,039	0,043	0,044	0,004	9,7
DBT = dibenzothiophene	0,009	0,008	0,008	0,008	0,008	0,001	8,8
4 - MDBT = Methyldibenzothiophene	0,022	0,019	0,017	0,018	0,019	0,002	9,9
2+3-MDBT = methyldibenzothiophene	0,009	0,009	0,007	0,008	0,008	0,001	9,5
1 - MDBT = methyldibenzothiophene	0,003	0,003	0,003	0,003	0,003	0,000	8,3

SDs = standard deviations.

The SDs of the AS crude were calculated from the RAs in Table 11 and the percentages of RSDs linked to these remained below 14 % (Table 12). These results can be extrapolated to the other crudes studied (Annex 1). Together with what is presented in Table 11, this provides support for the assertion that these hydrocarbons remained unchanged. That is to say, without appreciable changes in their chemical composition over the course of a year as established in the European standard CEN/TR 15522-2 defined by the European Committee for Standardisation in 2012.

Table 12. Diagnostic ratios obtained from AS crude oil.

RDs	AS1	AS2	AS3	AS4	Prom	SDs	RSDs
P/F	2,038	2,113	1,854	1,876	1,970	0,126	6,4
P/Π-C17	0,478	0,406	0,423	0,411	0,429	0,033	7,7
F/n-Ci8	0,239	0,223	0,239	0,245	0,236	0,010	4,1
n-C29/n-C17	0,545	0,527	0,566	0,591	0,557	0,028	5,0
H29/H30	0,873	0,858	0,766	0,804	0,825	0,050	6,0
10 . G30/G30 . H30	0,117	0,108	0,118	0,113	0,114	0,004	3,8
M30/H30	0,264	0,266	0,271	0,323	0,281	0,028	10,0
% S27	30,027	29,823	34,197	35,073	32,280	2,744	8,5
% S28	13,040	11,002	13,181	12,025	12,312	1,014	8,2
% S29	56,934	59,175	52,622	52,902	55,408	3,192	5,8
D27/S27	0,792	0,835	0,663	0,640	0,733	0,096	13,1
IMP	0,811	0,929	0,880	0,946	0,895	0,061	6,8
Rc	0,886	0,951	0,924	0,960	0,930	0,033	3,6
% 4-MeDBT	64,070	62,236	63,358	62,455	63,030	0,847	1,3
% 2+3-MeDBT	26,610	28,275	26,247	28,664	27,449	1,198	4,4
% 1-MeDBT	9,320	9,489	10,395	8,881	9,521	0,636	6,7
DBT/Ph	0,093	0,088	0,104	0,091	0,094	0,007	7,4

Averages = means, SDs = standard deviations, RSDs = relative standard deviations.

P/F = pristane/phytane, P/n-C_{17} = pristane/heptadecane, FZa-C_{18} = phytane/octadecane, n-C /n-C_{2917} = nonacosane/heptadecane, H29/H30 = hopane C_{29} /hopane C_{30} , * 10 x G30/G30 x C_{30} = 10 . gammacerane / gammacerane . hopane C_{30} . M30/H30 = morethane/hopane C_{30} , % S_{27} = percentage of cholestanes, % S_{28} = percentage of ergostanes, % S_{29} = percentage of stigmastanes, D27/S27 = diasterane C_{27} / cholestane, IMP = methylphenanthrene index, Rc = calculated vitrinite reflectance, * % 4-MeDBT = % 4-methyldibenzothiophene, * % 2 + 3-MeDBT = % 2 + 3-methyldibenzothiophene, * % I-MeDBT = % 1-methyldibenzothiophene, DBT/Ph = dibenzothiophene/phenanthrene.

On the other hand, the RDs can also be used to differentiate one crude oil from another, whether they belong to the same reservoir or basin. Based on this premise, the results obtained for the RDs of each crude oil constituted a set of unique values that are presented in Tables 13 and 14. This constitutes the starting point to differentiate or not the 15 samples with which we worked for 12 months through the multivariate statistics developed below.

Table 13. Diagnostic relationships of crude oil samples A, B and C.

RDs	AC	AS	AN	BI	BM	BS	CI	CO
P/F	2,07 ± 0,2	1,97 ± 0,1	1,66 ± 0,1	2,02 ± 0,1	1,88 ± 0,0	1,90 ± 0,2	1,80 ± 0,1	1,67 ± 0,1
P/n-C17	0,35 ± 0,0	0,43 ± 0,0	2,59 ± 0,2	0,20 ± 0,0	0,20 ± 0,0	0,18 ± 0,0	0,44 ± 0,0	0,39 ± 0,0
F/n-C18	0,20 ±	0,24 ±	2,59 ±	0,11 ±	0,12 ±	0,11 ±	0,28 ±	0,27 ±

	0,0	0,0	0,2	0,0	0,0	0,0	0,0	0,0
n-C2√n-C17	0,50 ± 0,1	0,56 ± 0,0	0,00 ± 0,0	0,15 ± 0,0	0,13 ± 0,0	0,10 ± 0,0	0,17 ± 0,0	0,19 ± 0,0
H29/H30	0,86 ± 0,1	0,82 ± 0,0	0,81 ± 0,0	1,54 ± 0,1	1,11 ± 0,0	1,46 ± 0,1	0,93 ± 0,0	0,82 ± 0,1
10.G30/G30.C30	0,08 ± 0,0	0,11 ± 0,0	0,06 ± 0,0	0,22 ± 0,0	0,17 ± 0,0	0,20 ± 0,0	0,13 ± 0,0	0,15 ± 0,0
M30/H30	0,25 ± 0,0	0,28 ± 0,0	0,31 ± 0,0	0,00 ± 0,0	0,00 ± 0,0	0,00 ± 0,0	0,00 ± 0,0	0,00 ± 0,0
% S27	33,4 ± 2,2	32,3 ± 2,7	43,9 ± 2,6	32,8 ± 1,1	32,6 ± 1,7	31,8 ± 1,3	50,4 ± 0,4	44,6 ± 0,5
% S28	16,7 ± 0,8	12,3 ± 1,0	19,7 ± 1,2	37,2 ± 0,9	36,8 ± 0,1	37,1 ± 1,2	29,9 ± 0,6	32,0 ± 0,3
% S29	49,9 ± 2,7	55,4 ± 3,2	36,4 ± 1,7	29,9 ± 1,5	30,6 ± 1,8	31,0 ± 1,2	19,6 ± 1,0	23,4 ± 0,3
D27/S27	0,59 ± 0,0	0,73 ± 0,1	0,80 ± 0,1	0,00 ± 0,0	0,00 ± 0,0	0,00 ± 0,0	1,33 ± 0,1	1,36 ± 0,1
IMP	0,73 ± 0,0	0,89 ± 0,1	1,12 ± 0,0	1,02 ± 0,1	0,81 ± 0,0	0,80 ± 0,1	1,35 ± 0,0	1,22 ± 0,0
Rc	0,84 ± 0,0	0,93 ± 0,0	1,05 ± 0,0	1,00 ± 0,0	0,89 ± 0,0	0,88 ± 0,0	1,18 ± 0,0	1,11 ± 0,0
4- MeDBT	56,7 ± 1,5	63,0 ± 0,8	68,7 ± 2,0	50,2 ± 0,1	51,9 ± 0,1	46,1 ± 1,6	57,9 ± 0,4	54,5 ± 0,4
%2+3-MeDBT	29,4 ± 0,9	27,4 ± 1,2	22,2 ± 2,0	32,9 ± 1,4	31,3 ± 0,6	29,9 ± 2,3	30,7 ± 0,4	31,9 ± 0,2
1- MeDBT	13,9 ± 0,7	9,52 ± 0,6	9,01 ± 0,8	16,8 ± 1,3	16,8 ± 0,7	24,1 ± 1,5	11,4 ± 0,0	13,5 ± 0,4
DBT/Ph	0,06 ± 0,0	0,09 ± 0,0	0,01 ± 0,0	0,06 ± 0,0	0,11 ± 0,0	0,11 ± 0,0	0,09 ± 0,0	0,06 ± 0,0

Idem Table 12.

Table 14. Diagnostic relationships of crude oil samples D, E and F.

RDs	DB	DA	EB	EA	FI	FM	FS
P/F	1,63 ± 0,0	1,74 ± 0,1	2,74 ± 0,2	2,47 ± 0,1	1,86 ± 0,1	1,76 ± 0,0	1,94 ± 0,0
P/n-C17	0,69 ± 0,0	0,70 ± 0,0	0,57 ± 0,0	0,78 ± 0,0	0,66 ± 0,0	0,67 ± 0,0	0,70 ± 0,0
F/n-C18	0,46 ± 0,0	0,45 ± 0,0	0,25 ± 0,0	0,37 ± 0,0	0,40 ± 0,0	0,41 ± 0,0	0,39 ± 0,0
C29/C17	0,25 ± 0,0	0,24 ± 0,0	0,36 ± 0,0	0,38 ± 0,0	0,36 ± 0,0	0,38 ± 0,0	0,31 ± 0,0
H29/H30	0,00 ± 0,0	0,00 ± 0,0	0,81 ± 0,0	0,88 ± 0,0	0,44 ± 0,0	0,38 ± 0,0	0,35 ± 0,0
G30/C30	0,00 ± 0,0	0,00 ± 0,0	0,07 ± 0,0	0,08 ± 0,0	0,08 ± 0,0	0,09 ± 0,0	0,09 ± 0,0
M30/H30	0,00 ± 0,0	0,00 ± 0,0	0,34 ± 0,0	0,49 ± 0,0	0,80 ± 0,1	0,67 ± 0,0	0,67 ± 0,0
% S27	59,8 ± 1,9	57,1 ± 1,2	50,8 ± 0,6	50,8 ± 0,3	52,7 ± 1,6	51,2 ± 1,7	48,2 ± 0,7
% S28	20,7 ± 1,6	24,8 ± 1,1	22,1 ± 0,5	22,7 ± 0,2	21,5 ± 0,6	24,1 ± 0,4	24,7 ± 1,0
% S29	19,5 ± 0,4	18,1 ± 0,3	27,1 ± 1,1	26,5 ± 0,2	25,8 ± 1,2	24,6 ± 1,4	27,1 ± 0,3
D27/S27	1,20 ± 0,0	1,34 ± 0,1	1,19 ± 0,0	1,30 ± 0,0	1,35 ± 0,1	1,50 ± 0,0	1,52 ± 0,0
IMP	1,21 ± 0,1	1,24 ± 0,1	0,69 ± 0,0	0,67 ± 0,0	0,99 ± 0,0	0,94 ± 0,0	1,00 ± 0,0
Rc	1,10 ± 0,0	1,12 ± 0,1	0,82 ± 0,0	0,81 ± 0,0	0,98 ± 0,0	0,96 ± 0,0	0,99 ± 0,0
4-MeDBT	58,9 ± 1,2	64,6 ± 1,9	57,0 ± 1,4	58,2 ± 0,7	57,6 ± 1,2	57,8 ± 1,9	57,2 ± 0,6
2-MeDBT	40,0 ± 1,2	34,4 ± 1,9	31,3 ± 0,8	28,4 ± 0,0	34,4 ± 1,6	36,1 ± 2,0	34,1 ± 0,1
I-MeDBT	1,05 ± 0,0	1,01 ± 0,1	11,7 ± 0,5	13,4 ± 0,6	7,96 ± 0,6	6,06 ± 0,1	8,73 ± 0,7
DBT/Ph	0,21 ± 0,0	0,14 ± 0,0	0,11 ± 0,0	0,20 ± 0,0	0,27 ± 0,0	0,28 ± 0,0	0,08 ± 0,0

Idem Table 12. * Their formulas are expressed in reduced formulas for a better visualisation of the table.

3.1.1 Multivariate analysis.

To evaluate the differences and similarities between the crude samples, the 16 geochemical parameters presented in Tables 13 and 14 were used. On the one hand, a classical cluster analysis was performed (Figure 11) in which it was firstly observed that the samples associated with each well remained constant throughout the study. Secondly, it is worth noting that the distribution of the crude oils was related to the geological formation to which they belong. In other words, two large groups were formed, each containing crudes from the

Springhill Fm and those from the Lower Magallanes Fm, which were well differentiated. It is important to note that the samples extracted from the Campo Indio reservoir (D) were different from the rest of the Springhill Fm crudes, probably because they were the only ones obtained from a 3,000 m deep well. Finally, it is important to note that the samples from the AN oil well in the Del Mosquito field showed the greatest differences from the rest of the hydrocarbons studied. Studies of this nature have been widely used to determine hydrocarbon families. Rangel et al. (2017) worked with hierarchical cluster analysis in Colombian crude oil samples to generate oil families from diagnostic relationships. Shorter distances in the obtained dendogram were associated with closer genetic relationships. In this regard, Li et al. (2022) consider that cluster analysis is an effective method for classifying genetic types of multi-source oils, and the key is to select effective oil-source correlation parameters, i.e. the appropriate diagnostic indices. In their study, they were able to generate a cluster with two major groupings or families of crude oils from the Junggar Basin in northwest China. In addition, they effectively determined the origin of the organic matter in the crude oils, the sedimentary water environment to which they were exposed, and their degree of thermal evolution.

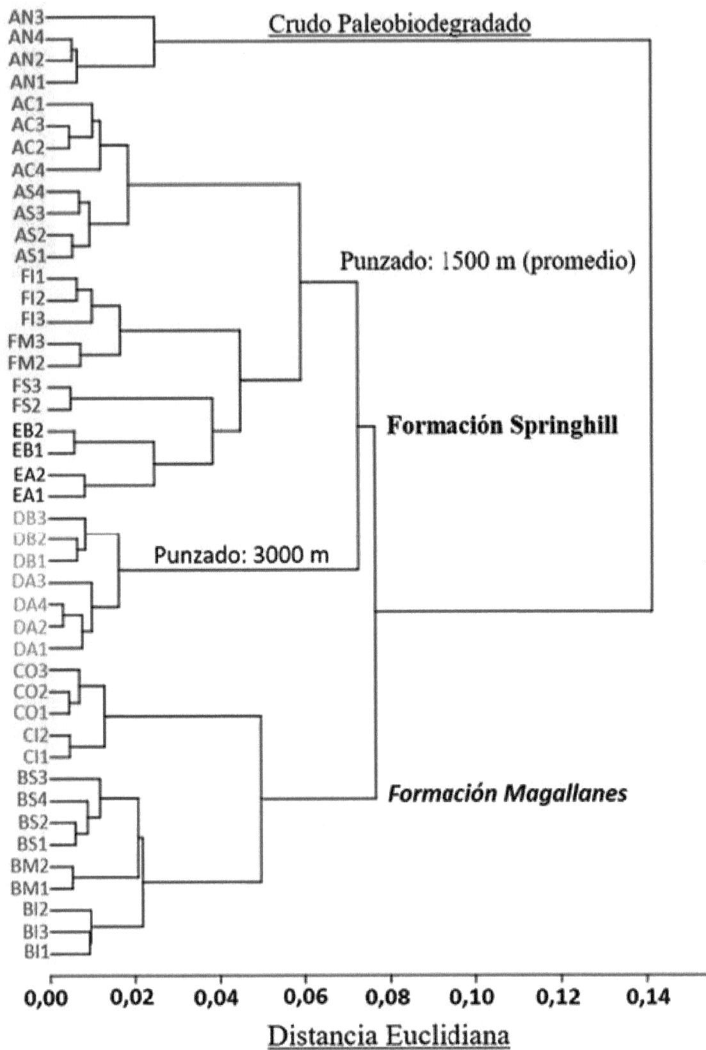

Figure 11. Dendogram of the 15 crude samples analysed during one year.

To complement the results presented in the dendogram (Figure 11), a three-dimensional PCA was carried out (Figure 12). This generated the separation of the reservoirs along the octants formed by the intersection of the principal component (PC), secondary component (SC) and tertiary component (TC). Interestingly, it is the CS, which accounts for approximately 28% of the analysis, that separated the crude oils by formation. The CS (+) contained the Springhill samples and the CS (-) the Lower Magellan samples. The cluster analysis showed a clear separation between the crudes from the Campo Indio (D) reservoir and the rest of the Springhill Fm oils. Furthermore, the PCA supported these differences from the CT, as the Campo Indio samples were located in the octant (CP+; CS+; CT-) far separated from the rest of the crudes belonging to its geological formation.

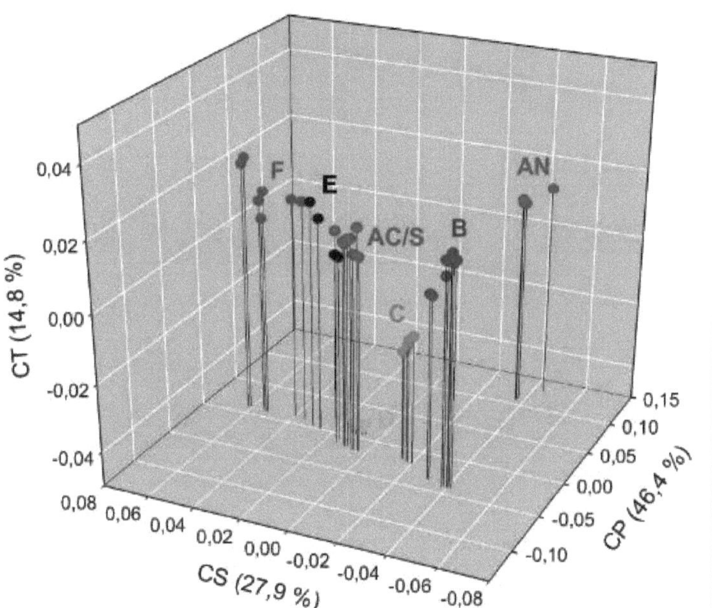

Figure 12. Principal Component Analysis of the 15 crude oil samples analysed during one year. A. Del Mosquito (red), B. Agua Fresca (blue), C. María Inés (green), D. Campo Indio (yellow), E. Cañadón Salto (black), F. La Maggie (violet).

2.5 . Geochemical characterisation.

3.2.1 Chromatograms.

From the TIC fraction defined in section 1.1.2, the n-alkanes described in Table 3 and, in addition, the acyclic isoprenoids pristane and phytane were determined by GC/MS for the 15 crude oils in Table 9 through the corresponding TICs. For the practical purposes of this work, only the TICs of six crude oil samples from one of the samples are presented in Figure 13. It should be noted that the distribution of n-alkanes in the crude oil samples AN, BS, CI, DB, EA and FI were characterised by a different pattern in each case. Firstly, AN (Del Mosquito - Springhill Fm) presented notable differences with respect to the rest, as a marked loss of linear alkanes and a prominent baseline uplift were observed. On the other hand, BS (Agua Fresca - Lower Magallanes Fm) exhibited a regular decreasing behaviour of its n-paraffins and EA (Cañadón Salto - Springhill Fm) resembled it, however, from n-C18 onwards the decrease in relative abundance was not as abrupt. Regarding the crude oil sample CI (María

31

Inés - Lower Magallanes Fm), it can be said that the decrease in n-alkane abundance as a function of the increase in carbon chain was not as marked as in the cases described above. Finally, both DB (Campo Indio - Springhill Fm) and FI (La Maggie - Springhill Fm) had a unimodal distribution of linear alkanes with a maximum between n-C16 and n-C17, and then experienced a marked decrease of higher molecular weight molecules. It is important to add that the chromatograms of the rest of the crudes that are not presented (AC, AS, BM, BI, CO, DA, EB, FM and FS) had an n-alkane profile similar to the crudes of the oilfields to which they belong. The only exception was the AN sample, which not only differed from AC and AS, but also from the rest of the hydrocarbons analysed.

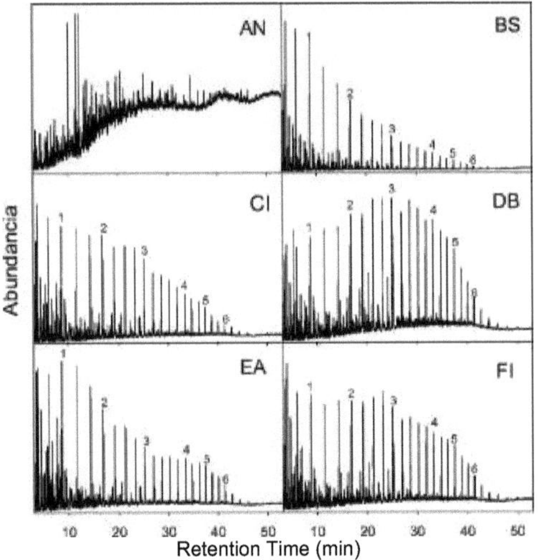

Figure 13. TICs of AN, BS, CI, DB, EA and FI crudes. 1: decane (n-C10), 2: tridecane (n-C13), 3: heptadecane (n-C17), 4: docosane (n-C22), 5: pentacosane (n-C25), 6: octacosane (n-C28).

The fragmentograms associated with the m/z = 191 ratio were obtained from the aliphatic fraction and are represented in Figure 14, where the tricyclic terpanes are observed on the left and the pentacyclic terpanes on the right (Table 4). From these fragmentograms, it was visualised that the AN crude (Del Mosquito - Springhill Fm) was the only one that presented clear and intense signals of these biomarkers with predominant peaks for hopanes H29 and H30. The EA (Cañadón Salto - Springhill Fm) and FI (La Maggie - Springhill Fm) samples showed low intensity peaks for both tricyclic and pentacyclic terpanes. On the other hand, in the BS (Agua Fresca - Lower Magallanes Fm) and CI (María Inés - Lower Magallanes Fm) crudes, hopanes H29 and H30 were slightly detectable. Finally, DB (Campo Indio - Springhill Fm) showed the tricyclic terpanes T23, T24 and T26. The missing m/z = 191 ion fragmentograms (AC, AS, BM, BI, CO, DA, EB, FM and FS) exhibited a very strong similarity between crudes from the same reservoir across the terpane profile. In this case, AN crude oil was related to AC and AS, and these preliminary results show the potential capacity of biomarkers to be used as chemical fingerprints of hydrocarbons. This is because the terpane profile showed virtually no significant changes between these hydrocarbons extracted from the Del Mosquito field.

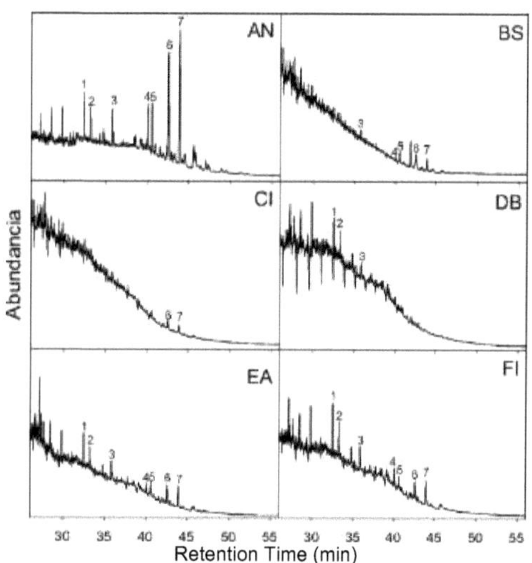

Figure 14. Fragmentograms (m/z = 191) of the crude oils AN, BS, CI, DB, EA and FI. 1: tricyclic terpane C_{23} (T23), 2: tricyclic terpane C_{24} (T24), 3: tricyclic terpane C_{26} (T26), 4: trisnorneohopane (Ts), 5: trisnorhopane (Tm), 6: norhopane C_{29} (H29), 7: hopane C_{30} (H30).

Fragmentograms associated with m/z = 217 were also obtained from the saturated hydrocarbon fraction, which corresponds to a family of tetracyclic biomarkers without aromaticity (Figure 15). Consistent with what was mentioned above for the terpanes, again the AN crude was the sample that allowed the best visualisation of these biomarkers derived from eukaryotic organisms. In addition, a similar distribution was observed for the Springhill Fm crudes (AN, DB, EA and FI), which were characterised by a strong presence of pregnane (peaks 1, 2 and 3), C_{27} diasteranes (4) and cholestanes (5). In contrast, the CI sample linked to the Lower Magallanes Fm presented with low intensity the biomarkers mentioned for the Springhill Fm. It should be noted that in the BS crude the differences were even greater because low abundances of C_{27} diasteranes (4), cholestanes (5), ergostanes (6) and stigmastanes (7) were evidenced. For practical purposes the m/z = 217 ion fragmentograms for AC, AS, BM, BI, CO, DA, EB, FM and FS were not presented because they well preserved this biomarker profile between crudes from the same reservoir. The results for AN, AC and AS can be perfectly extrapolated to those mentioned for terpanes, suggesting the great potential of these molecules to match crudes that have undergone palaeobiodegradation in the reservoir.

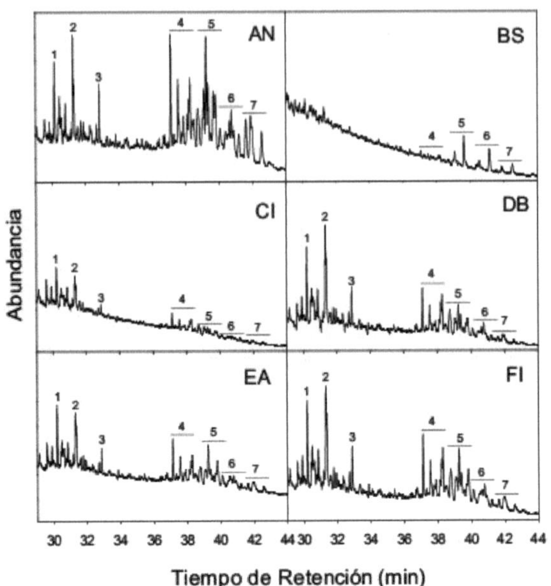

Retention Time (min)

Figure 15. Fragmentograms (m/z = 217) of the crude AN, BS, CI, DB, EA and FI. 1: C_{20} sterane ($s20$), 2: C_{21} sterane ($s21$), 3: C_{22} sterane ($s22$), 4: C_{27} diasteranes ($D27$). 5: cholestanes ($s27$), 6: ergostanes ($s28$), 7: stigmastanes ($s29$).

3.2.2 Precursor organic matter.

Using the $P/n\text{-}C17$ ratio as a function of $F/n\text{-}C18$ with the values presented in Tables 13 and 14, the Shanmugam Plot (Figure 16) was plotted to assess the precursor organic matter of the crude oils. Most of the samples were grouped in the lower left quadrant close to each other, in the zone defined as mixed organic matter. However, some crude oils did not exhibit this behaviour and departed to a lesser or greater extent from what was described above. On the one hand, hydrocarbons from the Agua Fresca - Lower Magallanes Fm reservoir (blue) were positioned in the mixed region, but away from the lower left corner of the graph, which highlights a high thermal maturity reached by the crudes positioned in that part of the graph. On the other hand, samples from the María Inés reservoir - Lower Magallanes Fm (yellow) were located at the boundary of the terrigenous-mixed interface. Finally, the most noticeable difference was the AN crude oil, which moved to the upper right of the Shanmugam diagram (Shanmugam 1985).

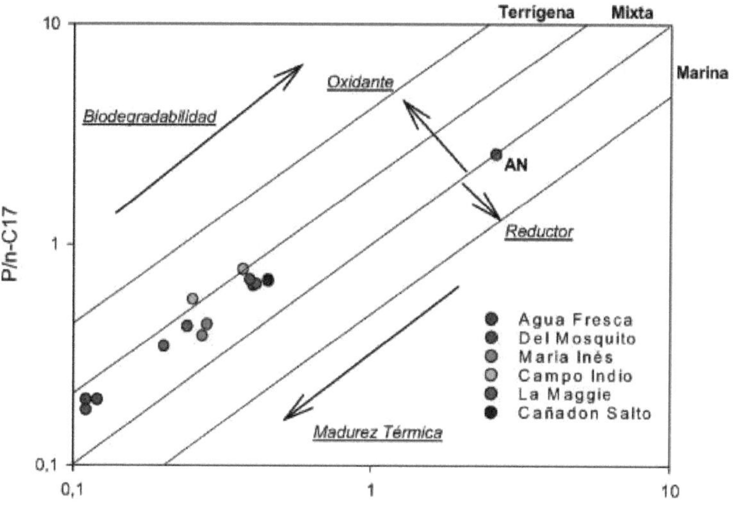

Figure 16. P/n-C17 as a function of F/n-C18 for the crude oils studied (Shanmugam 1985).

Another way to analyse the nature of the organic matter that settled and gave rise to these crudes by subsequent biological, physical and geochemical transformations is through the P/n-C17 ratio as a function of P/F (Figure 17), which complemented the results presented in the Shanmugam diagram. In this case, all the crudes were located in the range defined for mixed type matter. However, the AN crude oil was significantly distanced from the rest at the top of the graph, which is linked to paleobiodegradation alterations as already observed in Figure 16.

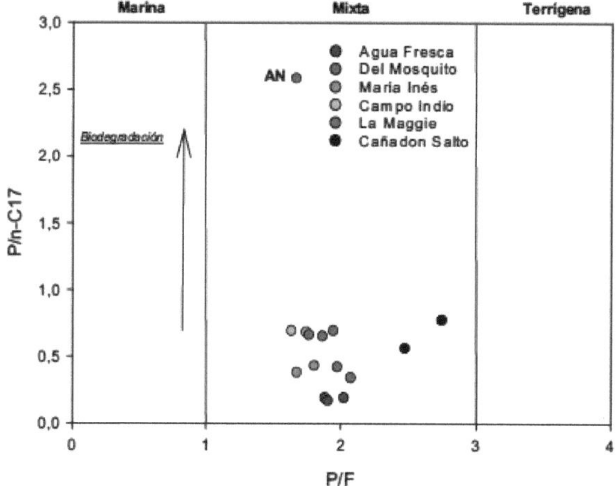

Figure 17. P/n-C17 as a function of P/F for the crude oils studied (Peters et al. 2005).

3.2.3 Source rock lithology.

From the relative percentages of the methylated dibenzothiophene isomers presented in Tables 13 and 14 it was possible to determine the lithology of the Palermo Aike and Margas Verdes formations (source rocks) based on the distribution of these PAHs (Hughes et al. 1995). The behaviour of these aromatic sulphur compounds followed a staircase-like pattern, i.e. the relative percentages decreased from 4-MeDBT to 2+3-MeDBT to 1-MeDBT in all cases, characteristic of a siliciclastic lithology.

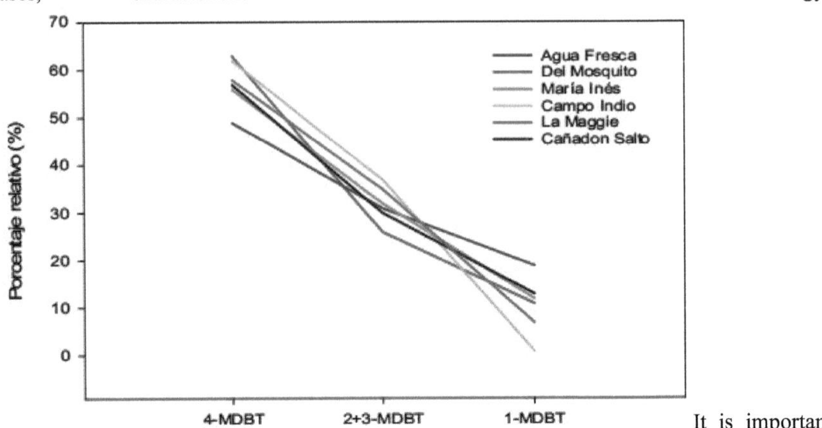

It is important to clarify that for a better visualisation of the trend lines, the average of the relative percentages of each well linked to its respective reservoir was taken, because the representation of the 15 samples made the graph unclear (Figure 18).

Figure 18. Percentages of MeDBTs for each of the reservoirs (Killops and Killops 2005).

Based on the proportions of steranes S_{27}, S_{28} and S_{29} (Tables 13 and 14), the ternary diagram proposed by Moldowan et al. (1985) was plotted. It showed that the samples were associated with a marine shale-type lithology, except in the case of the crude oil from the Del Mosquito reservoir (Figure 19). In these samples, a marine shale-type lithology was also observed, but calcareous.

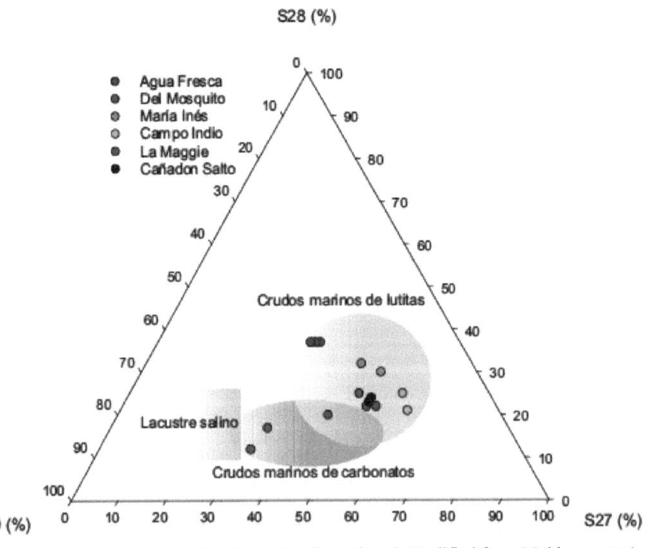

Figure 19. Ternary sterane diagram for the crude oils analysed. Modified from Moldowan et al. 1985.

The lithology present during the sedimentation of organic matter is a key aspect of oil formation. To understand this, the dibenzothiophene/phenanthrene (DBT/Ph) ratio was determined and displayed as a function of the P/F ratio (Figure 19). Hughes et al. (1995) proposed that the DBT/Ph ratio has the ability to assess the availability of reduced sulphur that is incorporated into organic matter.

It is also a good indicator of carbonate source rocks when the ratio is greater than one and of siliciclastic source rocks when the ratio is less than one.

On the other hand, the P/F ratio indicates the oxidoreductive conditions of the depositional environment. In this sense, the results allowed us to infer that the lithology of the Palermo Aike and Margas Verdes formations (source rocks) was siliciclastic (marine shale). It should be noted that all the crude oils were formed from organic matter that sedimented in marine environments, in which the oxidoreductive conditions were suboxic-detoxic.

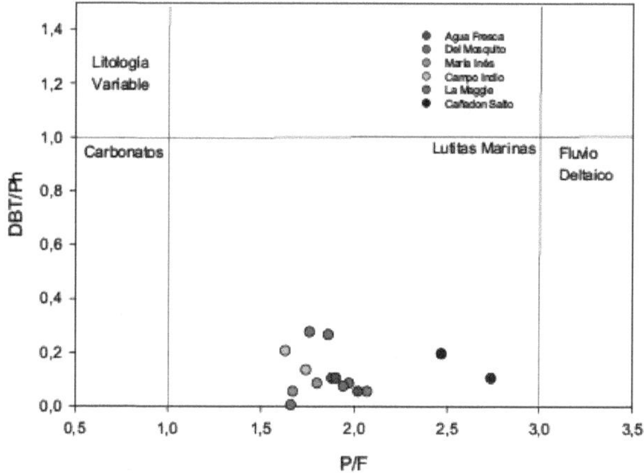

Figure 20. DBT/Ph vs P/F for the crude oils studied (Hughes et al. 1985).

In summary, the parameters evaluated suggest that the precursor organic matter of the 15 crude oils studied was of a mixed type, i.e. marine with continental input. Furthermore, the crude oils did not show palaeobiodegradation except in the case of the samples extracted from the AN oil well. On the other hand, the Palermo Aike Fm and the Margas Verdes Fm (source rocks) that generated these hydrocarbons are of a silicilastic marine nature (type shale).

3.3 Environmental stability.

Visualising the TICs is the starting point to determine whether there were changes in the composition of the crude oil samples under the laboratory conditions to which they were subjected (Figure 21). For the crude oil samples extracted from the seawater systems at time zero (T0), a bimodal distribution of n-alkanes was observed (Figure 21A), starting with nonane (n-C9) and ending with triacontane (n-C30). Between the first (T1) and second (T2) week, a decrease in the lighter n-alkanes characterised by a total decrease in n-C9 began to manifest itself (Figures 21B-C) and, on the other hand, there was a lifting of the baseline. During the third week (T3) and the first month (T4), the rise of the baseline became more marked (Figures 21D-E) and the distribution of n-alkanes was characterised by a decrease in the abundance of the following compounds: n-C10, n-C11, n-C12 and n-C13. At two months (T5) the n-alkanes up to tridecane (n-C13) decreased completely in abundance and the MCNR or hump in the chromatogram, which is typical evidence of biodegradation, increased (Figure 21F). This MCNR is made up of biodegradation products that cannot be separated by chromatography, thus constituting an unknown in the TIC (Choi et al. 2020). During T6, T7 and T8 (four, six and nine months, respectively), the crude oil showed a loss of n-C14, n-C15 and n-C16 molecules, and the baseline lifting became even more pronounced (Figures 21G-I). Finally, when the experiment was one year old (T9), a decrease in n-C17 and P was visualised in the TIC (Figure 21J), leading to a sharp decrease in the P/F ratio.

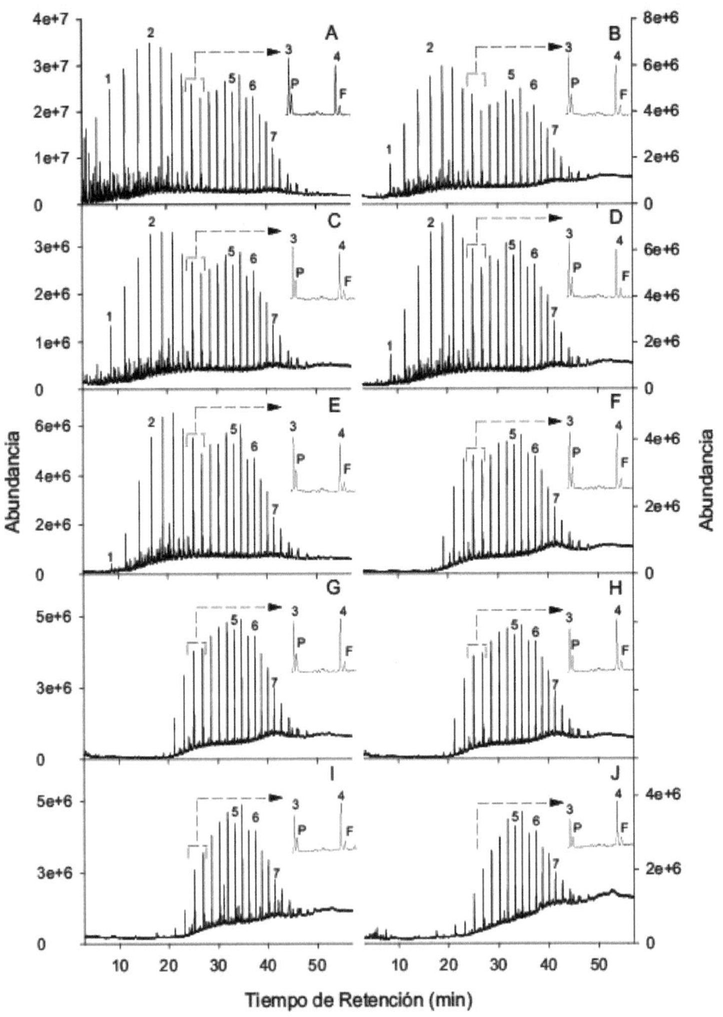

Figura 21.   TICs of AS crude oil extracted from seawater at times T0 (A) to T9 (J). 1: decane (n-C10), 2: tridecane (n-C13), 3: heptadecane (n-C17), 4: octadecane (n-C18), 5: docosane (n-C22), 6: pentacosane (n-C25), 7: octacosane (n-C28), P: pristane, F: phytane.

The crude oil samples that interacted with the soil were characterised by a similar distribution to that described for the seawater test. However, initially (T0) lower abundances of the n-alkanes between n-C9 and $n\text{-}C_{12}$ were evident and, in addition, baseline lifting occurred early in the test (Figure 22A). These differences were due to the crude oil handling process, which was necessary to achieve a homogeneous distribution of crude oil in the soil samples used in the experiment. During the first month (T1 to T4) there were no major changes in the pattern and abundance of n-alkanes, but there was a marked increase in the baseline (Figures 22B-E). After two months (T5), the loss of $n\text{-}C_{11}$ and $n\text{-}C_{12}$ was evident (Figure 22F) and, in addition, the most pronounced rise of the baseline at this time resulted during T6 (four months) in the appearance of a mixture of unresolved compounds (MCNR). This MCNR (Figure 22G) is associated with biodegradation-generated molecules that cannot be resolved under the analytical conditions (Sutton et al. 2005). On the other hand, the disappearance of n-alkanes to tetradecane ($n\text{-}C_{14}$) took place. The MCNR evolved at six months (T7) and stabilised between 9 (T8) and 12 months (T9) without noticeably affecting the overall composition of the weathered crude oil sample (Figures 22H- J).

In addition, the fragmentogram corresponding to ion m/z = 191 (Figures 23A) allowed us to visualise clear and intense signals from the tricyclic (C_{23} and C_{26}), pentacyclic (hopanes C_{29} and C_{30}) and extended hopanes (homohopanes C_{31}, C_{32} and C_{33}) terpanes. Notably, no decrease in the signal linked to the homo-hopanes (C_{31} to C_{33}) was observed. This was also evident in the relative concentrations of the remaining cyclic isoprenoids over the year of study for the crude oil sample subjected to artificial weathering both in reactors containing seawater (Figures 23B-J), as well as in reactors containing soil. It is important to add that a gradual lifting of the baseline was visualised through the fragmentograms presented in Figures 23A-J. This affected in the later stages of the test the heights of the tricyclic terpanes, but not the RDs of these cyclic isoprenoids presented in Table 15 and 16. For the practical purposes of the work the fragmentograms for the soil-weathered crude samples were not presented due to a marked similarity with those obtained for the seawater test.

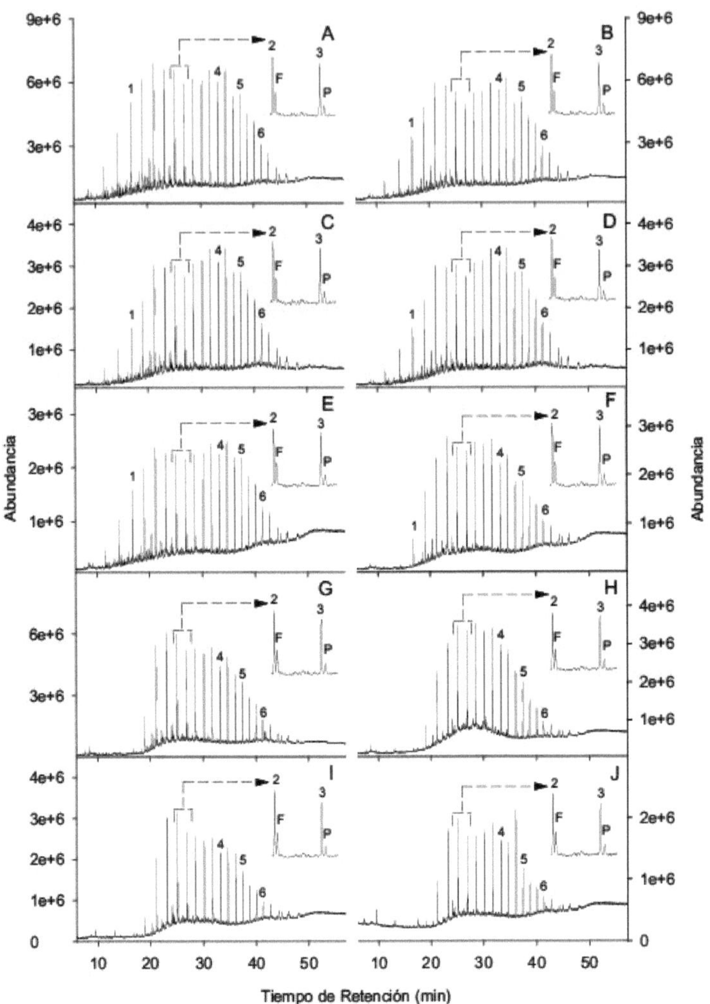

Figure 22. TICs of crude AS extracted from soil at times T0 (A) to T9 (J). 1: tridecane (n-C13), 2: heptadecane (n-C17), 3: octadecane (n-C18), 4: docosane (n-C22), 5: pentacosane (n-C25), 6: octacosane (n-C28), P: pristane, F: phytane.

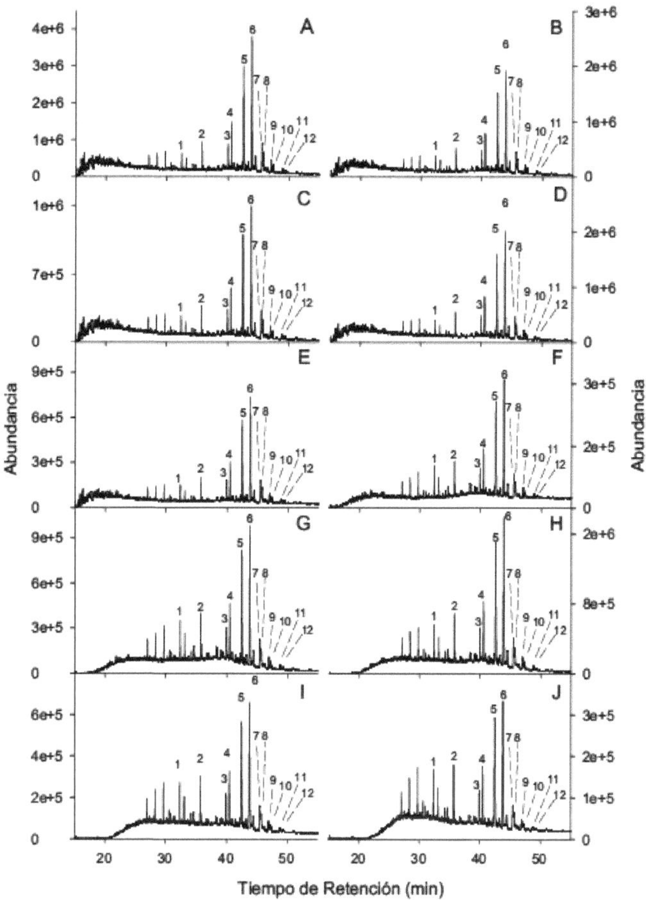

Retention Time (min)

Figura 22. Fragmentograms (m/z = 191) of AS crude extracted from seawater at times T0 (A) to T9 (J). 1: tricyclic terpane c_{23} (T_{23}), 2: tricyclic terpane c_{26} (T_{26}), 3: trisnorneohopane (Ts), 4: trisnorhopane (Tm), 5: norhopane c_{29} (H_{29}), 6: hopane c_{30} (H_{30}), 7-8: homohopanes c_{31} (R and S), 9-10: homohopanes c_{32} (R and S), 11-12: homohopanes c_{33} (R and S).

Finally, in the mass fragmentograms for ion m/z = 217, obtained from the crude samples exposed to the seawater reactors, a decrease in the relative abundances of the steranes (c_{27}, c_{28} and c_{29}) and of the c_{27} diasteranes as a function of the c_{21} pregnane and c_{22} homopregnane was observed. However, the relative intensity of the distribution pattern of the regular steranes (c_{27}, c_{28}, c_{29}) did not alter over the course of the experiment (Figures 24A-J). Furthermore, the steranes did not show a specific order of alteration after 365 days of weathering, however, the intensity of the signals between c_{21} and c_{22} with respect to c_{27}, c_{28} and c_{29} suggests a possible alteration by biodegradation (López and Infante 2021).

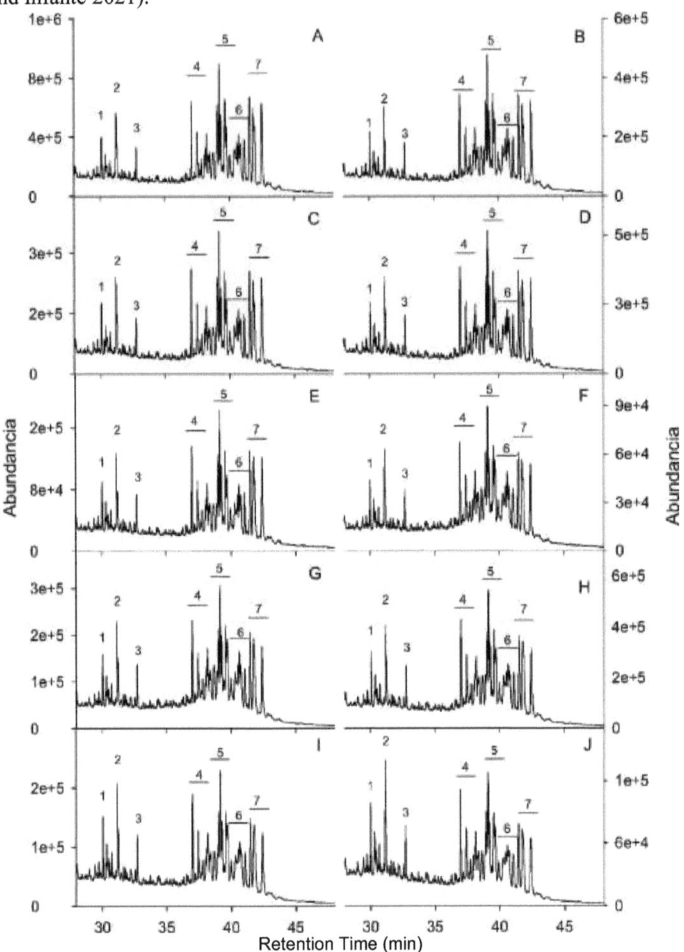

Figure 24. Fragmentograms (m/z = 217) of crude AS extracted from seawater at times T0 (A) to T9 (J). 1: c_{20} sterane (s_{20}), 2: c_{21} sterane (s_{21}), 3: c_{22} sterane (s_{22}), 4: c_{27} diasteranes (D_{27}). 5: cholestanes (s_{27}), 6: ergostanes (s_{28}), 7: stigmastanes (s_{29}).

Finally, it should be noted that these fragmentograms were practically identical for crude oils exposed to reactors with soil, therefore, only the results associated with hydrocarbons in aqueous systems were presented.

3.3.1 Environmental stability of crude oil in seawater.

The averages and SDs of the RDs for the seawater weathered crude oil test allowed the determination of the RSDs for each diagnostic index in the one-year interval (Tables 15 and 16). The analysis derived from this data matrix showed that the lowest RSD values were linked to the RDs from terpanes and steranes. Furthermore, ratios linked to the alicyclic isoprenoids pristane and phytane with their respective n-alkanes and to the methylated aromatic molecules associated with phenanthrene and dibenzothiophene were also below 5 %. On the other hand, only three ratios had an RSD higher than 5 %: P/F, (n-C13 + n-C14) / (n-C25 + n-C26) and (N0 + N1) / N2. These changes occurred over the time defined for the study, which was one year. It is interesting to note that two months (T5) after the start of the experiment, both (n-C13 + n-C14) / (n-C25 + n-C26) and (N0 + N1) / N2 decreased substantially. Finally, the P/F ratio did so after one year (T9), demonstrating greater stability. However, the use of these RDs is not suitable for analyses based on the comparison of suspect and environmental samples for spills occurring in seawater with residence times of 12 months or longer.

Table 15. Diagnostic relationships for AS crude oil in seawater at times T0 - T5.

RDs	T0	T1	T2	T3	T4	T5
P/F	$1,85 \pm 0,2$	$1,86 \pm 0,0$	$1,87 \pm 0,2$	$1,88 \pm 0,1$	$1,89 \pm 0,1$	$1,89 \pm 0,0$
P/n-C17	$0,41 \pm 0,0$	$0,39 \pm 0,0$	$0,43 \pm 0,0$	$0,42 \pm 0,0$	$0,43 \pm 0,0$	$0,43 \pm 0,0$
F/n-C18	$0,24 \pm 0,0$	$0,23 \pm 0,0$	$0,24 \pm 0,0$	$0,24 \pm 0,0$	$0,23 \pm 0,0$	$0,24 \pm 0,0$
(C13+C14)/(C25+C26)	$1,39 \pm 0,0$	$1,29 \pm 0,1$	$1,29 \pm 0,0$	$1,25 \pm 0,1$	$1,13 \pm 0,1$	$0,15 \pm 0,0$
(N0 + N1)/N2	$0,54 \pm 0,0$	$0,41 \pm 0,0$	$0,40 \pm 0,0$	$0,34 \pm 0,0$	$0,31 \pm 0,0$	$0,00 \pm 0,0$
2-MP/1-MP	$1,54 \pm 0,0$	$1,62 \pm 0,1$	$1,54 \pm 0,1$	$1,52 \pm 0,0$	$1,53 \pm 0,0$	$1,51 \pm 0,1$
4/1-MeDBT	$6,03 \pm 0,1$	$6,01 \pm 0,1$	$6,01 \pm 0,1$	$6,04 \pm 0,3$	$6,04 \pm 0,1$	$6,08 \pm 0,4$
2 + 3/1-MeDBT	$2,94 \pm 0,1$	$2,92 \pm 0,2$	$2,91 \pm 0,2$	$2,99 \pm 0,3$	$2,88 \pm 0,1$	$2,92 \pm 0,2$
Ts/H30	$0,10 \pm 0,0$	$0,11 \pm 0,0$	$0,11 \pm 0,0$	$0,11 \pm 0,0$	$0,11 \pm 0,0$	$0,12 \pm 0,0$
G30/H30	$0,03 \pm 0,0$	$0,03 \pm 0,0$	$0,03 \pm 0,0$	$0,03 \pm 0,0$	$0,03 \pm 0,0$	$0,03 \pm 0,0$
M30/H30	$0,08 \pm 0,0$	$0,08 \pm 0,0$	$0,08 \pm 0,0$	$0,08 \pm 0,0$	$0,08 \pm 0,0$	$0,08 \pm 0,0$
Ts/Tm	$0,49 \pm 0,0$	$0,50 \pm 0,0$	$0,49 \pm 0,0$	$0,49 \pm 0,0$	$0,50 \pm 0,0$	$0,51 \pm 0,0$
M30/H29	$0,18 \pm 0,0$	$0,17 \pm 0,0$	$0,16 \pm 0,0$	$0,17 \pm 0,0$	$0,17 \pm 0,0$	$0,16 \pm 0,0$
H31 (R)/H31 (S)	$0,57 \pm 0,0$	$0,58 \pm 0,0$	$0,58 \pm 0,0$	$0,58 \pm 0,0$	$0,57 \pm 0,0$	$0,55 \pm 0,0$
D27 βα (R)/H30	$0,03 \pm 0,0$	$0,03 \pm 0,0$	$0,03 \pm 0,0$	$0,04 \pm 0,0$	$0,04 \pm 0,0$	$0,04 \pm 0,0$
D27 β (S)/ D27 β (R)	$2,34 \pm 0,0$	$2,35 \pm 0,1$	$2,33 \pm 0,0$	$2,32 \pm 0,0$	$2,36 \pm 0,0$	$2,32 \pm 0,1$
S28 αββ (R + S)/H30	$0,05 \pm 0,0$	$0,06 \pm 0,0$	$0,06 \pm 0,0$	$0,06 \pm 0,0$	$0,06 \pm 0,0$	$0,07 \pm 0,0$
D27 β (R)/S29 a (S)	$0,20 \pm 0,0$	$0,21 \pm 0,0$	$0,20 \pm 0,0$	$0,22 \pm 0,0$	$0,23 \pm 0,0$	$0,24 \pm 0,0$
S29 aaa (S)/H30	$0,14 \pm 0,0$	$0,15 \pm 0,0$	$0,16 \pm 0,0$	$0,16 \pm 0,0$	$0,15 \pm 0,0$	$0,16 \pm 0,0$
S29α(S)/S29α(R+S)	$2,30 \pm 0,1$	$2,28 \pm 0,2$	$2,30 \pm 0,1$	$2,21 \pm 0,1$	$2,24 \pm 0,1$	$2,24 \pm 0,1$

Idem Table 16.

Table 16. Diagnostic relationships for AS crude oil in seawater at times T6 - T9.

RDs	T6	T7	T8	T9	RSDs
P/F	$1,84 \pm 0,2$	$1,84 \pm 0,1$	$1,83 \pm 0,1$	$1,06 \pm 0,0$	14,3 %
P/n-C17	$0,42 \pm 0,0$	$0,43 \pm 0,0$	$0,43 \pm 0,0$	$0,44 \pm 0,0$	3,26 %
F/n-C18	$0,22 \pm 0,0$	$0,22 \pm 0,0$	$0,22 \pm 0,0$	$0,24 \pm 0,0$	3,97 %
(C13+C14)/(C25+C26)	$0,01 \pm 0,0$	$0,00 \pm 0,0$	$0,00 \pm 0,0$	$0,00 \pm 0,0$	99,9 %
(N0 + N1)/N2	$0,00 \pm 0,0$	$0,00 \pm 0,0$	$0,00 \pm 0,0$	$0,00 \pm 0,0$	99,9 %
2-MP/1-MP	$1,50 \pm 0,0$	$1,47 \pm 0,0$	$1,46 \pm 0,0$	$1,46 \pm 0,0$	3,11 %
4/1-MeDBT	$6,10 \pm 0,2$	$6,04 \pm 0,2$	$6,04 \pm 0,0$	$6,03 \pm 0,1$	0,47 %
2 + 3/1-MeDBT	$2,94 \pm 0,0$	$2,98 \pm 0,1$	$3,00 \pm 0,1$	$2,94 \pm 0,2$	1,30 %
Ts/H30	$0,11 \pm 0,0$	$0,11 \pm 0,0$	$0,12 \pm 0,0$	$0,12 \pm 0,0$	3,01 %
G30/H30	$0,02 \pm 0,0$	$0,02 \pm 0,0$	$0,02 \pm 0,0$	$0,02 \pm 0,0$	4,46 %
M30/H30	$0,08 \pm 0,0$	$0,08 \pm 0,0$	$0,08 \pm 0,0$	$0,08 \pm 0,0$	2,75 %
Ts/Tm	$0,50 \pm 0,0$	$0,50 \pm 0,0$	$0,50 \pm 0,0$	$0,52 \pm 0,0$	1,97 %
M30/H29	$0,16 \pm 0,0$	$0,16 \pm 0,0$	$0,16 \pm 0,0$	$0,16 \pm 0,0$	4,10 %
H31 (R)/H31 (S)	$0,55 \pm 0,0$	$0,54 \pm 0,0$	$0,57 \pm 0,0$	$0,57 \pm 0,0$	2,35 %
D27 βα (R)/H30	$0,04 \pm 0,0$	$0,03 \pm 0,0$	$0,03 \pm 0,0$	$0,04 \pm 0,0$	4,20 %
D27 β (S)/ D27 β (R)	$2,34 \pm 0,0$	$2,30 \pm 0,0$	$2,35 \pm 0,1$	$2,32 \pm 0,0$	0,75 %
S28 αββ (R + S)/H30	$0,07 \pm 0,0$	$0,05 \pm 0,0$	$0,04 \pm 0,0$	$0,04 \pm 0,0$	4,55 %

D27 β (R)/S29 α (S)	0,24 ± 0,0	0,30 ± 0,0	0,28 ± 0,0	0,26 ± 0,0	4,63 %
S29 ααα (S)/H3o	0,15 ± 0,0	0,14 ± 0,0	0,14 ± 0,0	0,16 ± 0,0	4,50 %
S29α(SyS29α(R+S)	2,26 ± 0,2	2,24 ± 0,1	2,23 ± 0,1	2,20 ± 0,1	1,46 %

P/F = pristane/phytane, P/n-C_{17} = pristane/heptadecane, F/n-C18 = phytane/octadecane, (C_{13} + C_M)(C_{25} + C_{26}) = (tridecane + tetradecane) / (pentadecane + hexadecane), (N_0 + N_1)/N2 = (naphthalene + methylnaphthalene) / dimethylnaphthalene, 2-MP/1-MP = 2/1-methylphenanthrene, 4/1- MeDBT = 4/1-methyldibenzothiophene, 2 + 3/1/1-MeDBT = 2 + 3/1-methyldibenzothiophene, Ts/H_{30} = trisnorneohopane/hopane C_{30} , G_{30} /H_{30} = gamma-hopane/hopane C_{30} , M_{30} /H_{30} = morethane/hopane C_{30} , Ts/Tm = trisnorneohopane/trisnorneohopane, M_{30}/H_{29} = morethane/hopane C_{29} , H_{31} (R)/H_{31} (S) = homohopanes, D_{27} βα (R)/H_{30} = diasterane/hopane C_{30} , D_{27} β (S)/ D_{27} β (R) = diasteranos, S_{28} αβββ (R + S)/H_{30} = ergostanos/hopano C_{30} , D_{27} β (R)/S_{29} α (S) = diasterano/stigmastano, S_{29} αααα (S)/H_{30} = stigmastano/hopano C_{30} , S_{29} α(S)/S_{29} α(R+S) = stigmastanos.

3.3.2 Environmental stability of crude oil in soil.

The RSD values for crude oil disseminated in soil remained below 5 % for most cases (Tables 17 and 18), however, the samples have weathered under laboratory conditions after one year. This is because some ratios exceeded the permitted limit, which is indicative of weathering processes on the crude oil. The RSDs exhibited for the ratios P/n-C17, F/n-C18, (n-C13 + n-C14) / (n-C25 + n-C26) had values of approximately 7, 8 and 60 %, respectively. This behaviour was due to the loss of the light alkanes n-C13, n-C14, n-C17 and n-C18. The naphthalene ratio showed a gradual decrease in its average values and not an abrupt drop as its peer developed in section 3.2.2. On the other hand, the P/F ratio has remained below the permitted limit and can be considered together with the rest of the ratios as a potential tool to solve problems related to the accidental or intentional spillage of crude oil in soil for time intervals of 12 months.

Table 17. Diagnostic relationships for AS crude oil in soil at times T0 - T5.

RDs	T0	T1	T2	T3	T4	T5
P/F	2,08 ± 0,0	2,03 ± 0,2	2,04 ± 0,1	2,05 ± 0,0	2,10 ± 0,2	2,01 ± 0,0
P/n-C17	0,41 ± 0,0	0,41 ± 0,0	0,41 ± 0,0	0,41 ± 0,0	0,41 ± 0,0	0,43 ± 0,0
F/n-C18	0,23 ± 0,0	0,23 ± 0,0	0,25 ± 0,0	0,23 ± 0,0	0,24 ± 0,0	0,23 ± 0,0
(C13+C14)/(C25+C26)	0,95 ± 0,1	0,81 ± 0,0	0,83 ± 0,1	0,83 ± 0,0	0,84 ± 0,0	0,82 ± 0,1
(No + N1)/N2	0,20 ± 0,0	0,19 ± 0,0	0,18 ± 0,0	0,17 ± 0,0	0,16 ± 0,0	0,15 ± 0,0
2-MP/1-MP	1,58 ± 0,1	1,58 ± 0,0	1,62 ± 0,0	1,59 ± 0,0	1,58 ± 0,0	1,52 ± 0,0
4/1-MeDBT	6,10 ± 0,3	6,02 ± 0,1	6,02 ± 0,1	6,06 ± 0,1	6,08 ± 0,1	5,98 ± 0,1
2 + 3/1-MeDBT	2,73 ± 0,2	2,71 ± 0,2	2,70 ± 0,0	2,70 ± 0,1	2,70 ± 0,1	2,71 ± 0,2
Ts/H30	0,11 ± 0,0	0,11 ± 0,0	0,10 ± 0,0	0,11 ± 0,0	0,11 ± 0,0	0,10 ± 0,0
G30/H30	0,03 ± 0,0	0,03 ± 0,0	0,03 ± 0,0	0,03 ± 0,0	0,02 ± 0,0	0,03 ± 0,0
M30/H30	0,08 ± 0,0	0,08 ± 0,0	0,09 ± 0,0	0,08 ± 0,0	0,08 ± 0,0	0,08 ± 0,0
Ts/Tm	0,49 ± 0,0	0,49 ± 0,0	0,49 ± 0,0	0,49 ± 0,0	0,48 ± 0,0	0,48 ± 0,0
M30/H29	0,17 ± 0,0	0,16 ± 0,0	0,16 ± 0,0	0,16 ± 0,0	0,15 ± 0,0	0,17 ± 0,0
H31 (R)/H31 (S)	0,58 ± 0,0	0,57 ± 0,0	0,56 ± 0,0	0,56 ± 0,0	0,54 ± 0,0	0,54 ± 0,0
D27 βα (R)/H3o	0,03 ± 0,0	0,03 ± 0,0	0,03 ± 0,0	0,04 ± 0,0	0,04 ± 0,0	0,03 ± 0,0
D27 β (S)/ D27 β (R)	2,33 ± 0,1	2,41 ± 0,1	2,35 ± 0,1	2,39 ± 0,0	2,42 ± 0,1	2,33 ± 0,0
S28 αββ (R + S)/H3o	0,06 ± 0,0	0,06 ± 0,0	0,06 ± 0,0	0,06 ± 0,0	0,06 ± 0,0	0,05 ± 0,0
D27 β (R)/S29 a (S)	0,22 ± 0,0	0,22 ± 0,0	0,22 ± 0,0	0,23 ± 0,0	0,24 ± 0,0	0,23 ± 0,0
S29 aaa (S)/H3o	0,15 ± 0,0	0,15 ± 0,0	0,15 ± 0,0	0,16 ± 0,0	0,16 ± 0,0	0,14 ± 0,0
S29α(SyS29α(R+S)	2,21 ± 0,1	2,22 ± 0,1	2,18 ± 0,1	2,27 ± 0,0	2,25 ± 0,1	2,23 ± 0,0

Idem Table 16.

Table 18. Diagnostic relationships for AS crude oil in soil at times T6 - T9.

RDs	T6	T7	T8	T9	RSDs
P/F	2,11 ± 0,0	1,99 ± 0,1	2,06 ± 0,1	1,98 ± 0,0	2,07 %
P/n-C17	0,46 ± 0,0	0,46 ± 0,0	0,48 ± 0,0	0,49 ± 0,0	7,61 %
F/n-C18	0,22 ± 0,0	0,22 ± 0,0	0,28 ± 0,0	0,27 ± 0,0	8,24 %
(C13+C14)/(C25+C26)	0,40 ± 0,0	0,22 ± 0,0	0,09 ± 0,0	0,07 ± 0,0	59,6 %

$(N_0 + N_1)/N_2$	$0,12 \pm 0,1$	$0,09 \pm 0,0$	$0,06 \pm 0,1$	$0,05 \pm 0,1$	39,6 %
2-MP/1-MP	$1,59 \pm 0,1$	$1,61 \pm 0,0$	$1,62 \pm 0,1$	$1,64 \pm 0,1$	2,11 %
4/1-MeDBT	$6,01 \pm 0,1$	$6,02 \pm 0,2$	$6,04 \pm 0,0$	$6,00 \pm 0,3$	0,62 %
2 + 3/1-MeDBT	$2,72 \pm 0,0$	$2,70 \pm 0,0$	$2,72 \pm 0,1$	$2,78 \pm 0,3$	0,96 %
Ts/H_{30}	$0,11 \pm 0,0$	$0,11 \pm 0,0$	$0,11 \pm 0,0$	$0,11 \pm 0,0$	3,57 %
G_{30}/H_{30}	$0,03 \pm 0,0$	$0,02 \pm 0,0$	$0,02 \pm 0,0$	$0,02 \pm 0,0$	4,77 %
M_{30}/H_{30}	$0,08 \pm 0,0$	$0,08 \pm 0,0$	$0,08 \pm 0,0$	$0,08 \pm 0,0$	3,40 %
Ts/Tm	$0,49 \pm 0,0$	$0,49 \pm 0,0$	$0,50 \pm 0,0$	$0,52 \pm 0,0$	2,48 %
M_{30}/H_{29}	$0,16 \pm 0,0$	$0,17 \pm 0,0$	$0,15 \pm 0,0$	$0,15 \pm 0,0$	4,96 %
H_{31} **(R)**$/H_{31}$ **(S)**	$0,53 \pm 0,0$	$0,56 \pm 0,0$	$0,55 \pm 0,0$	$0,52 \pm 0,0$	3,67 %
D_{27} **βα (R)**$/H_{3o}$	$0,03 \pm 0,0$	$0,03 \pm 0,0$	$0,03 \pm 0,0$	$0,03 \pm 0,0$	3,83 %
D_{27} β **(S)**$/ D_{27}$ β **(R)**	$2,31 \pm 0,1$	$2,31 \pm 0,1$	$2,35 \pm 0,1$	$2,35 \pm 0,1$	1,64 %
S_{28} **αββ (R + S)**$/H_{3o}$	$0,05 \pm 0,0$	$0,05 \pm 0,0$	$0,05 \pm 0,0$	$0,05 \pm 0,0$	4,60 %
D_{27} β **(R)**$/S_{29}$ **a (S)**	$0,22 \pm 0,0$	$0,23 \pm 0,0$	$0,24 \pm 0,0$	$0,22 \pm 0,0$	3,80 %
S_{29} **aaa (S)**$/H_{3o}$	$0,15 \pm 0,0$	$0,14 \pm 0,0$	$0,14 \pm 0,0$	$0,14 \pm 0,0$	3,68 %
$S_{29}a(S_yS_{29}a$**(R+S)**	$2,24 \pm 0,1$	$2,12 \pm 0,0$	$2,23 \pm 0,1$	$2,23 \pm 0,1$	1,85 %

Idem Table 16.

3.3.3 Histograms of crude oil in seawater.

The relative percentages of PAHs over the time defined for the crude oil in seawater study varied (Figure 25). Initially in the T0 aqueous system (Figure 25A) the presence of naphthalene (N_0) and its methylderivatives, phenanthrene (P_0) and methylphenanthrenes (P_1), dibenzothiophene (D_0) and methyldibenzothiophenes (D_1) was observed. The major compounds were dimethylnaphthalenes (N_2) with a relative abundance of less than 30 %. Between T1 and T4 (Figures 25B-E) there was a total loss of naphthalene and a decrease of methylnaphthalenes (N_1), and dimethylnaphthalenes (N_2) were outnumbered in proportion by trimethylnaphthalenes (N_3). Between two and four months (T5 and T6), the histograms (Figures 25F-G) showed first the disappearance of methylnaphthalenes and then of dimethylnaphthalenes, a marked decrease in trimethylnaphthalenes and a significant increase in tetramethylnaphthalenes (N_4), in phenanthrene and mainly in methylphenanthrenes to become the dominant PAHs with relative percentages higher than 40 %. After six and nine months (T7 and T8) the chromatograms showed a decrease of tri- and tetramethylnaphthalenes and a marked stability of phenanthrene and methylphenanthrenes (P_1) as the most abundant with percentages around 60 % for the latter group (Figures 25H-I). Finally, after one year, tri- and tetramethylnaphthalenes decreased considerably, phenanthrene remained almost unchanged and methylphenanthrenes continued their trend with values above 60 % (Figure 25J). It is noteworthy that dibenzothiophene (D_0) remained almost unchanged over the 12 months with minimal relative concentrations and that methyldibenzothiophenes (D_1) showed a slight increase over the same period.

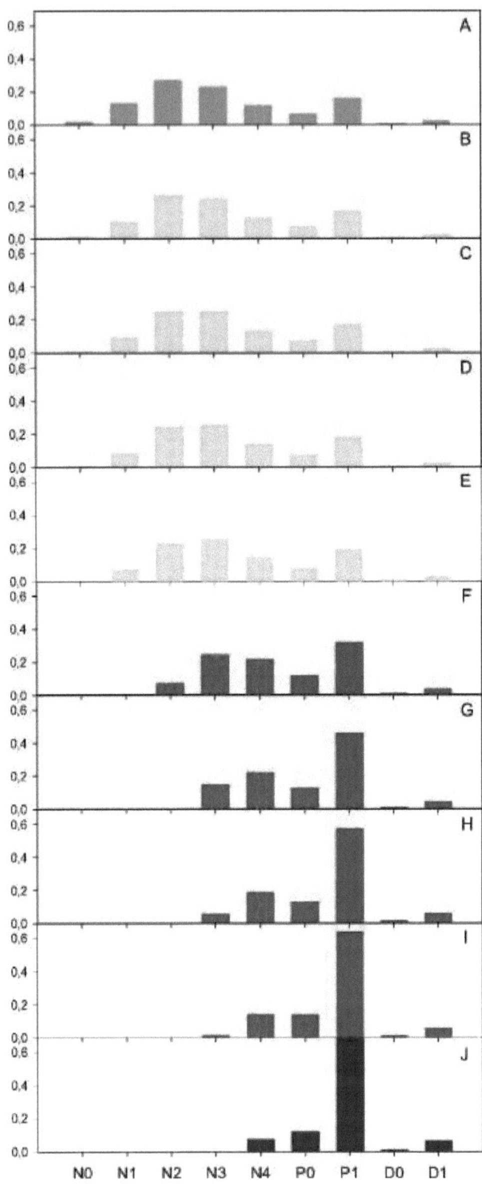

Polycyclic Aromatic Hydrocarbons

Figure 25. Histograms of PAHs for crude oils extracted from seawater at times T0 (A) to T9 (J). Nx: naphthalene and methylderivatives, Px: phenanthrene and methylderivatives, Dx: dibenzothiophene and methylderivatives.

3.3.4 Histograms of crude oil in soil.

The relative percentages of PAHs varied in each of the stages of this study (Figure 26). Initially in the soil systems (Figure 26A), the presence of naphthalene (N0) and its methylderivatives (N1, N2, N3, N4), phenanthrene (P0) and methylphenanthrenes (P1), dibenzothiophene (D0) and methyldibenzothiophenes (D1) was observed. The major compounds were trimethylnaphthalenes (N3) with a relative abundance of less than 30 %. Between T1 and T4 (Figures 26B-E) there was a sustained loss of naphthalene and its methylderivatives that can be removed from the soil surface mainly by evaporation, microbial oxidation and desorption (An et al. 2005). Between two and four months (T5 and T6), a steady decrease in naphthalene and its methylated derivatives was evident in the histograms (Figures 26F-G). In addition, there was a significant increase in phenanthrene and mainly in methylphenanthrenes to become the major PAHs with relative percentages around 40 %. After six and nine months (Figures 26H-I) the histograms showed the continuous decrease of naphthalene and its methylmolecules, a more prominent growth of phenanthrene and the consolidation of methylphenanthrenes (P1) as the most abundant with percentages close to 50 %. Finally, after one year (Figure 25J), naphthalene was almost gone, its methylated derivatives greatly diminished in relation to the original crude. Phenanthrene and methylphenanthrenes continued to increase their relative abundances to values close to 20 and 60 %, respectively. With regard to the sulphur PAHs, it can be mentioned that dibenzothiophene (D0) remained practically unchanged over the 12 months with a low relative abundance. Methyldibenzothiophenes (D1) underwent a gradual decrease over the same period from 3 to 2 % (Figures 26A-J).

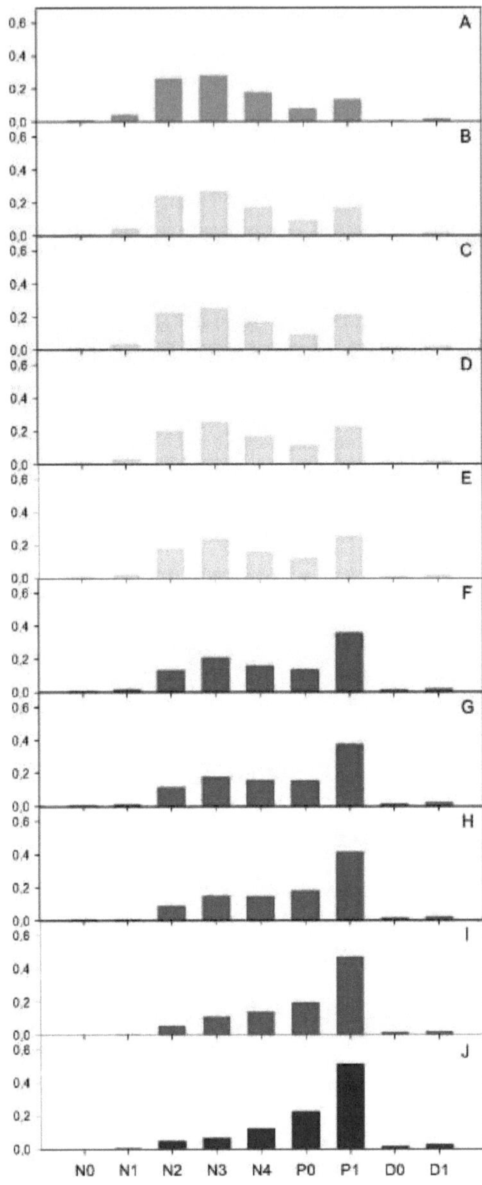

Polycyclic Aromatic Hydrocarbons

Figure 26. Histograms of PAHs for crude oils extracted from soil at times T0 (A) to T9 (J). NX: naphthalene and methylderivatives, PX: phenanthrene and methylderivatives, DX: dibenzothiophene and methylderivatives.

4 DISCUSSION

This section discusses the results obtained during the course of this research. Initially, the reproducibility of the 15 crude oils from samples that were taken over a period of one year is evaluated, and then the degree of relatedness between them is analysed using multivariate statistics. Subsequently, the chemical characterisation of these hydrocarbons is carried out by means of their n-alkane profiles, PAHs, biomarkers and the DRs derived from these molecules. In the second part, the stability of n-alkanes, PAHs and biomarkers from AS crude oil against weathering is studied. This experiment takes place under laboratory conditions for 12 months after the crude oil has been poured into seawater reactors on the one hand and soil reactors on the other.

4.1 . Study of the 15 crude oils from the Southern Basin.

4.1.1 Reproducibility of crude oils.

The RSDs obtained from each analysed molecule (section 3.1.2) and the RDs derived from them (section 3.1.3) for the 15 crudes in this study showed values below 14 % in their corresponding samples. Therefore, the results determined that the crude oil samples extracted for both the Springhill Fm and Lower Magallanes Fm reservoirs were stable over the course of a year. It is worth noting that this behaviour was expected, as within the geological scale it constitutes a short period of time (Peters et al. 2005; López and Lo Mónaco 2017). In other words, within reservoirs, hydrocarbons are subjected to various processes of a physical, chemical and/or biological nature that alter their chemical composition when geological time is significant. That is, when the time span is such that phenomena such as biodegradation, gravitational segregation, water washing, thermal alteration, among others, take place (Killops and Killops 2005; López et al. 2019). On the other hand, that RSDs have had percentages between 1 and 13 % is within the bias associated with the analytical variability of the methodologies employed, the complex nature of the samples and the various compartments with which it is in contact from reservoir to uptake (Zhang et al. 2015).

4.1.2 Multivariate statistical analysis.

The cluster analysis (section 3.1.4) allowed a marked distribution of the crude oil samples by levels to be visualised. In the first instance, the hydrocarbons were grouped in relation to the oil well from which they were extracted. It is important to mention that we are talking about the same crudes and that, in addition, this correlates with the results obtained for stability as a function of time developed in the previous section. Secondly, the samples used during this study were linked to their respective reservoirs. This is associated with the contained geological environments that these petroleum reservoirs present and that generally allow the chemical composition of hydrocarbons to be preserved (Peters et al. 2005; Lorenzo et al. 2018). The third level was determined by the depth of puncture from which each of the oils were captured. This parameter generated a separation of the Springhill Fm hydrocarbons into two groups. On the one hand, those reservoirs with a depth of extraction of 1500 m (Del Mosquito, Cañadón Salto and La Maggie) and on the other hand, only the Campo Indio reservoir with drilling below the surface of 3000 m. Furthermore, principal component analysis supported these results, as the Campo Indio samples were linked to the TC (-) and the rest of the hydrocarbons belonging to its geological formation were associated with the TC (+). The greater depths with respect to the land surface are accompanied by increases in temperature, in accordance with the thermal gradient of the basin. This heat promotes a higher thermal maturity of oils located in deeper regions and explains the differences found

(Hardebol et al. 2009; Han et al. 2019).

Within a basin, geological formations represent the most differential point at the time of contrasting crudes coming from them, as they are rock bodies characterised by common lithological properties that differentiate them from adjacent ones (Cunningham and Mann 2007). Based on this premise, it is reasonable to understand why the dendogram is separated into two main groups. The first one is composed of practically all the crudes belonging to the Springhill Fm (AC, AS, DA, DB, EA, EB, EB, FI, FM and FS) and the second one is composed of the hydrocarbons extracted from the Lower Magallanes Fm (BI, BM, BS, CI and CO). In this sense, it is interesting to note that in the principal component analysis it was the CS that separated the crude oils by formation, on the one hand, the CS (+) contained the samples extracted from Springhill and the CS (-) those from Lower Magallanes. This situation can be explained by the source rocks that generated these hydrocarbons. Numerous papers have published the correlation between the Palermo Aike Fm. petroleum and the Springhill reservoir unit (Pittion and Gouadain 1992; Pittion and Arbe 1999; Rodriguez et al. 2008). On the other hand, some petrological rocks of the Lower Magallanes reservoir rock could be correlated with the Margas Verdes Fm (Villar and Arbe 1993). Furthermore, according to Robbiano and Arbe (1996), petrological rocks from the Campo Boleadoras and Maria Inés reservoirs (Lower Magallanes Fm) are also correlatable with the Margas Verdes Fm. This means that correlations have been established in the Austral Basin that allow a general recognition of the relationship between source rocks and discovered hydrocarbons. However, correlations often present a certain degree of uncertainty, given the similarity between the source sections of the Palermo Aike/Inoceramus Inferior, Margas Verdes, and lateral equivalents (Rodriguez et al. 2008).

Finally, only the samples obtained from a single oil well did not respond to the general behaviour mentioned above. In the case of AN crude oil, a marked separation from the rest of the hydrocarbons analysed in this study was observed in both the dendogram (Figure 17) and the PCA (Figure 18). Regarding the latter, it is interesting to mention that all crude oils were located in a narrow range along the COP with values between -0.05 to 0.05, however, AN was positioned at COP values higher than 0.10. This situation could be explained by the analysis of its TIC (Figure 14), which showed a marked loss of linear alkanes and a prominent baseline rise leading to the formation of a Complex Unresolved Mixture (MCNR). The MCNR, or hump in the chromatogram, is typical evidence of biodegradation and this process takes place once the crude oil is stored in the reservoir (Lopez and Infante 2021). The AN hydrocarbon initially presented a similar composition to its peers AC and AS. However, the causes that triggered the activity of microorganisms to degrade its major components (n-alkanes) from a phenomenon known as paleobiodegradation, which developed over millions of years, are unknown (López and Infante 2021). The profile of n-alkanes, PAHs, biomarkers and the DRs obtained in section 5.1.1 for each of the 15 crude oils together constitute a fingerprint of the crude oils. Furthermore, from the results obtained in section 5.1.2 we can state that these hydrocarbons are different from each other, therefore, we can identify them unequivocally through their corresponding fingerprint. Then, if we extrapolate this analysis to all the crudes extracted from the Austral Basin at present, we could generate a database that stores the fingerprint of each one of them.

4.1.3 Terpanes and steranes.

Terpane distributions obtained from ion m/z = 191 (Figure 15) revealed low concentrations of these biomarkers for hydrocarbons from the Agua Fresca and María Inés reservoirs

(Magallanes Fm), and Campo Indio, Cañadón Salto and La Maggie (Springhill Fm). This behaviour is characteristic of the saturated hydrocarbon fraction of condensates, covering the n-paraffin range between 7 and 11 carbon atoms (Schwarzkopf and Leythaeuser 1988; Orea et al. 2021). In other words, these samples behave like an atmospheric distillation cut-off in the gasoline range, and the concentrations of terpanes and steranes are often below the detection limit of the gas chromatograph used. However, the physical properties of the crudes under study (Agua Fresca, María Inés, Campo Indio, Cañadón Salto and La Maggie) do not correspond to those of a condensate. On the other hand, Rullkotter and Welte (1980) mention that this phenomenon is not only observed in condensates (Speight 2014). It is also exhibited by thermally mature crudes that have undergone high-temperature maturation (120 - 150 °C) in their storage rock, levels equivalent to vitrinite reflectance values (% Ro) of 1.6 (Dow 1977; Han et al. 2019), and by tetra- and pentacyclic hydrocarbons that have undergone thermal cracking. But in crudes with the n-paraffin distribution reported in this study, the presence of tricyclic diterpanes should be observed in the fragmentogram for ion m/z = 191. Unless a high concentration of n-paraffins and acyclic branched alkanes in the saturated fraction has masked the presence of these biomarkers in the samples (Richardson and Miller 1982; Fang et al. 2019).

Regardless of the above, for the Lower Magallanes Fm crudes (Agua Fresca and María Inés), the distribution of steranes was obtained from the m/z = 217 ion (Figure 16), with a very low intensity. Only the presence of regular ßß steranes was observed, which are those that generally appear at the highest concentration (Seifert and Moldowan 1978; El-Sabagh et al. 2018). In these cases, the signals are weak, but identifiable by attributing this behaviour to what is described in the previous paragraph. On the other hand, the hydrocarbons from Campo Indio, Cañadón Salto and La Maggie (Springhill Fm) showed a more noticeable pattern of these biomarkers in which pregnanes stood out to the left of the fragmentograms as the most abundant compounds. However, there was an exception in the terpane and sterane profile (Figure 15 and 16) which corresponded to samples AC, AN and AS (Del Mosquito deposit - Springhill Fm). These crudes showed a clear and marked distribution of both groups of biomarkers, which may be related to their geological environment.

4.1.4 Type of organic matter.

The organic source of all the crudes studied was of a mixed, mainly marine type, with a minority continental contribution (Tissot and Welte 1984; López and Lo Mónaco 2010). This is suggested by the P/n-C17 and F/n-C18 ratios represented in the Shanmugam diagram (Figure 19). This idea is reinforced by the n-alkane distributions observed in the TICs in Figure 14 and the low values of the n-C29/n-C17 ratio (Tables 13 and 14). This trend has already been reported in the Austral Basin, as studies on crude oils extracted from the Springhill Fm also showed this nature of the kerogen (Rodriguez et al. 2008; Tomas et al. 2020). On the other hand, previous work by Villar and Arbe (1993) showed that the crudes extracted from the Lower Magallanes Fm (central-northern zone) and the source rock extracts from the Margas Verdes Fm were type II/III (mixed) kerogenic. These results were inferred from microscopic observations that were associated with an algal-amorphous material with a minor terrigenous mixture and biomarker fingerprints (similar to the fragmentograms obtained). In this sense, by means of a Rock-Eval pyrolysis analysis for the rocks, they validated the presence of mature type II/III kerogen. In addition, they defined the Margas Verdes Fm as the source rock for the crude oil extracted from the Lower Magallanes Fm.

(Cagnolatti and Miller 2002). Furthermore, it can be assured that no crude oil was produced in the intercalated pelites of the Tobifera Series (lacustrine source rock), since the kerogen generated in this facies is mainly terrigenous (type III) and in low proportion algal (type I; Legarreta and Villar 2011).

4.1.5 Palaeobiodegradation.

When analysing the composition of crude oil samples, it is important to consider whether changes associated with the phenomenon of palaeobiodegradation have occurred. This biological degradation of hydrocarbons mediated by micro-organisms in the reservoir over millions of years has a direct impact on the composition of the hydrocarbons. It also indirectly affects the characterisation of the type of precursor organic matter (Peters et al. 2005; López and Infante 2021). Of the 15 crude oils included in this research, 14 of them were characterised by the absence of desmethylhopanes in the fragmentograms for ion m/z = 177 (not presented in the paper). Furthermore, the $P/n\text{-}C_{17}$ and $F/n\text{-}C_{18}$ values below 0.8 (Tables 13 and 14), the profile of the n-alkanes observed in Figure 14 and the distribution of their points in Figures 19 and 20 indicate that these hydrocarbons have not been affected by paleobiodegradation. The petroleum in the Austral Basin has originated in Lower Cretaceous rocks, specifically from inputs of mainly marine organic matter, with secondary continental inputs. The analysis of the accumulated hydrocarbons in its reservoirs carried out by Cagnolatti et al. (1996) has indicated that no biodegradation processes have occurred in these reservoirs.

The only exception was linked to AN crude oil from the Del Mosquito field, which showed $P/n\text{-}C_{17}$ and $F/n\text{-}C_{18}$ ratios above 2 (Table 13). It was also characterised by an irregular and greatly decreased n-alkane pattern (Figure 14) and a positioning in the upper right zone of the Shanmugam diagram, far away from the rest of the crudes (Figure 19), inferring that a paleobiodegradation process has affected this hydrocarbon. Biological alteration processes have not been documented in the Austral Basin, but they have been a constant in the history of oil generation and accumulation in the Golfo San Jorge Basin. In more detail, Villar et al. (1996) documented the basis for these alteration processes in the Bajo Barreal Fm of the Golfo San Jorge Basin, concluding that the hydrocarbons studied resulted from complex biodegradation processes over long periods of geological time.

4.1.6 Sedimentation conditions and bedrock lithology.

The 1, 2 + 3 and 4-Methyldibenzothiophene (MeDBT) isomers provide information on the source rock lithology (Killops & Killops 2005; Abdulazeez & Fantke 2017). The staircase-like pattern of the relative percentages of these MeDBT isomers allowed us to propose that the nature of the source rocks is siliciclastic (Figure 21). This correlates with the literature referring on the one hand to the Margas Verdes Fm, which is described as a source rock consisting of shales and marine marls (Rodriguez et al. 2008). On the other hand, the rocks of the Palermo Aike Fm are black marine shales that are in direct contact with the sandy reservoirs of the Springhill Fm (Rodriguez et al. 2008).

The dibenzothiophene/phenanthrene (DBT/Ph) ratio as a function of pristane/phytane (P/F) is used as an indicator of the depositional environment of sedimentary rocks (Hughes et al. 1995; Rangel et al. 2017). The samples studied for the Austral Basin were located within the lower intermediate zone in the diagram in Figure 22. This suggests that the sedimentation conditions were associated with a marine shale palaeoenvironment, deposited under low oxygenation conditions given by P/F values around 2. A study carried out on crude oils extracted from the Del Mosquito reservoir belonging to the Springhill Fm of the Austral Basin obtained similar results that are associated with marine and suboxic depositional conditions (Tomas et al. 2020). In addition, Pittion and Goudaian (1992) proposed a sedimentary model in which the marine shales of the Palermo Aike Fm (source rock) were deposited in a low-oxygen water layer. In this low-oxygen environment, organic matter was preserved and allowed this sedimentary sequence to develop the initial potential to generate oil. Finally, Rodriguez et al. (2008) also proposed that the deposition of the Palermo Aike Fm. occurred under dysaerobic to anaerobic conditions.

4.2 Environmental stability of AS crude oil.

4.2.1 Chromatograms

A progressive loss of light n-alkanes and a continuous lifting of the baseline during the first month was observed in the crude samples that remained in the reactors with seawater (Figures 21A-E). This can be explained by a synergistic effect between the volatization of the light n-alkanes and an incipient biodegradation of the same. Samples of a crude oil subjected to weathering for a period of 30 days in the laboratory (Olson et al. 2017) using artificial (sterilised) seawater with the addition of nutrients and without/with the presence of a chemical dispersant underwent evaporation of their light n-alkanes (n-C10 to n-C13). However, the same crude oil under the same conditions but exposed to seawater presented n-C17 as the first compound in its n-alkane profile, suggesting biodegradation as the dominant process (Olson et al. 2017). Comparing these results with those obtained for hydrocarbons in lithospheric systems, some differences can be described (Figures 22A-E). Firstly, low molecular weight n-alkanes were retained longer in the study systems, probably adsorbed to the soil matrix that opposes their evaporation (Peters et al. 2005). Secondly, the emergence and evolution of baseline uplift was more marked due to the microbial community altering the hydrocarbons present (Sutton et al. 2005; Ali et al. 2020).

After two months (T5), n-alkanes up to tridecane (n-C13) had completely decreased in abundance in the crude oil samples over seawater (Figure 21F), but not in the hydrocarbons extracted from soil. For the latter, biodegradation and evaporation (Figure 22F) were considered to be responsible for the total decrease of the light n-alkanes undecane (n-C11) and dodecane (n-C12), and of tridecane (n-C13) only after four months (T6). In addition, the formation of a MCNR occurred as a result of the gradual and progressive increase of the baseline. Its presence is typical evidence of biodegradation (Figure 22G) and consists of organic compounds that cannot be identified under the analytical conditions used for the separation of n-alkanes through TICs (Lopez and Infante 2021). That is, it consists of bioresistant compounds such as cyclic saturates, aromatics, naphthoaromatics, and polar compounds that cannot be separated by chromatography technique (Sutton et al. 2005; Lundberg 2019). These behaviours were reflected in quantitative results from controlled laboratory simulations of weathered hydrocarbons. Relative decreases in lower molecular weight aliphatic molecules (n-C11 to n-C15) occurred in these simulations, but no significant decreases were observed after 15 weeks for n-alkanes > n-C15 (Agüero-Manzano 2019).

During T6, T7 and T8 (four, six and nine months, respectively) the crude oil in seawater evidenced a loss of n-C14, n-C15 and n-C16 molecules, and the baseline uplift became even more pronounced (Figures 21G-I). However, samples extracted from soil between T7 and T8 only showed the decrease of tetradecane (n-C14), but a noticeable increase of MCNR (Figures 22H-I). The lower molecular weight n-alkanes are the first to disappear when hydrocarbons are released into the environment. This is due to the vapour tension they present and to a rapid metabolisation of the microorganisms when they obtain energy from these molecules (López and Infante 2021). Finally, when the experiment was one year old (T9), a decrease in n-C17 and

P was visualised in the TIC (Figure 21J), which led to a sharp decrease in the P/F ratio (Table 19). These results suggest that evaporation and biodegradation coexisted as alteration phenomena in the crude oil samples tested in seawater. However, on the MCNR, high intensity signals are observed for the n-alkanes between n-C17 and n-C30, and the acyclic isoprenoids P and F, which indicates in principle that these compounds have not been biodegraded (López and Infante 2021). In the case of the samples contained in the reactors with soil, 12 months after the experiment, a decrease in the abundances of heptadecane (n-C17) and octadecane (n-C18) was observed with respect to the acyclic isoprenoids pristane (P) and phytane (F), respectively (Figure 22J). This behaviour is characteristic of a microorganism-mediated process that results in the degradation of linear alkanes that are easier to metabolise than P and F, which have a branched structure (Peters et al. 2005; Lobao et al. 2022). On the other hand, it is important to add that above the MCNR, high intensity signals were observed for n-alkanes between pentadecane (n-C15) and tridecane (n-C30), suggesting in the first instance that these molecules have not been biodegraded (Stout and Wang 2016). These results can be compared with those obtained in other studies. For example, one study assessed the presence and distribution of n-alkanes and aliphatic biomarkers in soil samples collected in Owaza, Niger Delta (Nigeria) over a one-year period, and found that the aliphatic fraction was dominated by n-alkanes from n-C19 to n-C31 (Faboya et al. 2016). Another investigation carried out in Amazonian soil linked to oil spills from facilities in that region showed a highly weathered state of the crude oils. The distribution of n-alkanes was characterised by the absence of their light components and by the predominance of high molecular weight compounds several years after the incident (Rosell-Melé et al. 2018).

4.2.2 Fragmentograms

This did not affect the distribution pattern presented by the terpanes (Figure 23A-J), where tricyclic (c23 and c26), pentacyclic (hopanes c29 and c30) and extended hopanes (homohopanes c31, c32 and c33) terpanes were detected. Notably, no decrease in the signal linked to the homohopanes (c31 to c33) or in the relative concentrations of the other cyclic isoprenoids was observed (Figure 23B-J). In oilfields, the biodegradation of these biomarkers takes place by the loss of a methyl group, giving rise to 25-norhopanes or demethylated hopanes (López and Infante 2021), which can be identified through the mass fragmentogram m/z = 177. For the samples of this study, these compounds were not observed, which reinforces the results obtained for the homohopanes (data not shown). On the other hand, in the mass fragmentograms for ion m/z = 217, a decrease in the relative abundances of the steranes (c27, c28 and c29) and of the c27 diasteranes as a function of the c21 pregnane and c22 homopregnane was observed. However, the relative intensity of the distribution pattern of the regular steranes (c27, c28, c29) did not alter during the experiment (Figure 24A-J). The steranes did not present a specific order of alteration after 365 days of weathering, however, the intensity of the signals between c21 and c22 with respect to c27, c28 and c29 suggests a possible alteration by biodegradation (López and Infante 2021).

Laboratory weathering tests on crude oils have shown that both terpanes and steranes were not affected by evaporation phenomena, as they concentrated proportionally with respect to the increasing percentages of degraded crude oils (Wang and Fingas 2003). These results show a relative stability of these cyclic isoprenoids under these test conditions. In addition, the recalcitrance to weathering of terpanes and steranes has been documented in numerous investigations. For example, a biodegradation study carried out in the laboratory showed that there were no signs of alteration in the composition of terpanes and steranes. The results were independent of crude oil type (light, medium or heavy), incubation times (7, 14 and 28 days), incubation conditions (4, 10, 15, and 22 °C), and presence or absence of nutrients (Swannel et al. 1996; Wang et al. 1998; Yang et al. 2023). Other studies associated with light crude oil spills in seawater that was enriched with nutrients exposed the pronounced stability of hopane H_{30}. This recalcitrance was exhibited against biodegradation for periods of six months at an average temperature of 15 °C in the laboratory (Prince et al. 1994; John et al. 2018) and 14 weeks in plots constructed on the beach (Venosa et al. 1997, Zhang et al. 2015). Furthermore, no photooxidation was observed in this biomarker when samples were irradiated with UV lamps for 48 hours at a distance of 15 cm in the laboratory (Garrett et al. 1998; Yang et al. 2016). Its biological degradation and that of other biomarkers has been achieved under aggressive laboratory conditions using aerobic enrichment cultures (Douglas et al. 2012; John et al. 2018).

On the other hand, a study carried out by Rosell-Melé et al. (2018) in the northern Peruvian Amazon made it possible to determine who was responsible for the oil pollution. This was due to the fact that the soil samples analysed maintained constant terpane and sterane profiles several years after the oil spill following the rupture of an oil pipeline. Another contingency of this nature occurred in Nigerian lithospheric systems and was studied 12 months after the oil spill. In this case, it became evident that the crude oil responsible for the contamination came from a field located in the Niger Delta due to the integrity that the distributions of these biomarkers maintained (Faboya et al. 2016). The ability of terpanes and steranes to indicate the source responsible for oil spilled in the open is due to their refractory nature and high resistance to biodegradation (Garcia et al. 2019). In addition, the study of a spill in the Gulf of Mexico after a year and a half showed a relative enrichment of these molecules with respect to alkanes (Aeppli et al. 2012). Finally, biological degradation of these biomarkers has been achieved under aggressive laboratory conditions using aerobic enrichment cultures (John et al. 2018). Alteration of these compounds has also been observed in several field studies involving time periods longer than five years (Lobao et al. 2022). For example, eight years after an experiment in which oil was spilled in a mangrove swamp in Guadeloupe (Lesser Antilles), total steranes and hopanes were reduced by more than 25 % (Reyes et al. 2014). Similar observations were made 20 years later from an oil spill in Antarctica (Rodriguez et al. 2018).

4.2.3 Alkane and isoprenoid RDs

The RDs presented in Tables 17 and 18 for the crude oil samples extracted from seawater and soil, respectively, presented RSDs values below 5 % in most cases. Therefore, these ratios are considered to be unaffected by the test conditions as published by Zhang et al. (2015). From the areas, $P/n\text{-}C_{17}$ and $F/n\text{-}C_{18}$ ratios were calculated to be used as biodegradation indices, as they compare compounds with different degrees of resistance. In this sense, pristane and phytane are more resistant to biodegradation with respect to $n\text{-}C_{17}$ and $n\text{-}C_{18}$ (López and Infante 2021). These $P/n\text{-}C_{17}$ and $F/n\text{-}C_{18}$ ratios did not exceed 3 % of the RSD for hydrocarbons in seawater. This indicates that preferential depletion of easily degradable molecules ($n\text{-}C_{17}$ and $n\text{-}C_{18}$) versus acyclic P and F isoprenoids has not occurred, which has also been observed in other studies such as those presented by Yim et al. (2011) and Zhang et al. (2015). On the other hand, if we consider evaporation as the predominant weathering phenomenon during the test, it can be better explained that the values of $P/n\text{-}C_{17}$ and $F/n\text{-}C_{18}$ remained relatively constant. This is because $n\text{-}C_{17}$, $n\text{-}C_{18}$, P and F compounds have similar vapour pressures and pass into the gaseous state with the same tendency (Turner et al. 2014; Orea et al. 2021). However, for the lithospheric systems both ratios (P/n-$C17$ and F/n-$C18$) exceeded 5 % of their RSDs, indicating that the crude oil samples in soil have been biodegraded. This behaviour has also been observed in numerous studies such as those presented by Aeppli et al. (2012), Faboya et al. (2016) and Rosell- Melé et al. (2018). On the other hand, the ratio (n-$C13$ + n-$C14$) / (n-$C25$ + n-$C26$) associated with the evaporation process exhibited an RSD value close to 60 % (Table 18). In detail, a progressive decrease in the RD value was observed as the test time elapsed. This situation can be explained by a higher evaporation rate of the n-$C13$ and n-$C14$ alkanes with respect to their n-$C25$ and n-$C26$ pairs. It is important to add that this situation was also reflected for the oil tested in the seawater reactors (Table 17), but in addition the P/F ratio was above the permitted limit. This can be interpreted from the lower molecular masses of F, n-$C13$ and n-$C14$, and hence their higher evaporation rates with respect to P, n-$C25$ and n-$C26$, respectively, which generated a decrease in their RDs affecting in turn their RSDs values.

The RDs determined for terpanes and steranes (Tables 17 and 18) were below 5 % of the RSD in all cases, suggesting that they have not been affected by evaporation and incipient biodegradation that have taken place during the test year. Regarding the former, when evaporation of a crude oil takes place, high molecular mass compounds such as these biomarkers tend to increase their relative concentration in the residual crude oil without modifying their DRs (Wang et al. 2006; López and Infante 2021). On the other hand, biodegradation has not taken long enough under the test conditions to pronouncedly affect the relative abundances of terpanes and steranes in both seawater and soil. The high stability of these cyclic isoprenoids was documented in investigations associated with environmental contaminations with crude oil or simulated in the laboratory over short time intervals (days) or very long periods (several years). For example, a study published by Zhang et al. (2015) that evaluated the 12 RDs of terpanes and steranes presented (Tables 17 and 18) showed that the ratios did not exceed 5 % of the RSD after three months of monitoring. On the other hand, responsibility for oil contamination of soils in the Niger Delta could be assigned through the determination of parameters such as Ts/Tm that did not change one year after the spills (Faboya et al. 2016). Furthermore, no change in the Ts/Tm ratio was observed 24 years after a spill in the Strait of Magellan, and 25 years after the Nipisi accident, the $H29/H30$ and Ts/Tm ratios were still stable (Song et al. 2016). Finally, the extent of environmental contamination from the Deepwater Horizon oil spill in the Gulf of Mexico could be determined from residual samples. After 10 years, the same terpanes and steranes that were initially found at the time of the accident were still present in these samples (Arekhi et al. 2021).

4.2.4 PAHs RDs and histograms

PAHs have been used to understand the origin and environmental fate of crude oil and its derivatives, as they are relatively stable molecules. For example, numerous investigations linked to the Exxon Valdez oil spill have found that some proportions of PAHs remain constant in early and intermediate stages of weathering (Douglas et al. 1996). One of the derived RDs for PAHs is (N0 + $N1$) / N2 which relates the concentrations between naphthalene and its methylderivatives, and in this work suggests that an evaporation process has affected the weathered crude oil samples. For the seawater test, from T0 to two months later (T5), the value decreased to zero (Table 17). In the case of the soil experiment, a slight and constant decrease in this ratio was observed over the course of the year, resulting in an RSD close to 40 % (Table 18). Therefore, this indicates in both cases that naphthalene and methylnaphthalenes have evaporated at a higher rate than dimethylnaphthalenes and this behaviour is associated with their lower molecular weight (Zhang et al. 2015). It is worth noting that the rest of the ratios linked to PAHs did not undergo substantial changes, i.e. they remained below 5 % of their RSD. For example, the 2-MP/1-MP (methylphenanthrene) ratio has the ability to indicate whether photooxidative processes have affected PAHs (Olson et al. 2017). The RSD of this ratio did not exceed 5 % and considering that the test reactors were placed throughout the experiment out of direct incidence of light, it is suggested that light has not promoted changes in the composition of the crude oil samples subjected to weathering in seawater and soil.

Along the same lines other RDs originating from PAHs have been used, both 4-MeDBT and 2 + 3-MeDBT (methyldibenzothiophenes) are less resistant to biological degradation than 1-MeDBT and are used to indicate the effects of biodegradation (Wang and Fingas 2003; Kao et al. 2015). The RSDs of 4-MeDBT/1- MeDBT and 2 + 3-MeDBT/1-MeDBT were lower than 4 %, suggesting that these ratios have not yet been affected by a biodegradation process considering the time and test conditions in both systems (Tables 17 and 18). Furthermore, it is important to mention that the soil microbial community was not stimulated by the use of nutrients. Research using heterotrophic microorganisms in a controlled laboratory microenvironment to study bioremediation in crude oil contaminated soils showed the same number of bacteria 55 days after the start of the trial (Mariano et al. 2007). This may be associated with a slowdown in biodegradation linked to a change in microbial function, rather than a reduction in bacterial population (Greenwood et al. 2008; Lopez and Infante 2021). However, there is a history of studies that showed significant changes in the DRs of PAHs. For example, samples taken over the course of eight months after the Hebei Spirit spill in South Korea showed a progressive decrease in both 4-MeDBT/1-MeDBT and 2 + 3-MeDBT/1-MeDBT due to microbially mediated degradation processes (Yim et al. 2011). In this regard, in a simulation of natural weathering of crude oil carried out in China by Zhang et al. (2015), in which an 8 m pond[3] was excavated at a distance of 50 m from the sea, seawater was pumped in, and then about 50 L of crude oil (average temperature = 15 °C) was poured in. After 95 days from the start of the test, the 4-MeDBT/1-MeDBT and 2- MP/1-MP ratios exceeded 5 % of their RSDs, indicating that both biodegradation and photolysis were acting on the crude oil components. Finally, in a set of chronic discharges into seawater in northeastern Brazil, it was observed that the RSDs described in this paragraph did not meet the standards established by CEN (2012). These contaminations, which were more than one year old, were exposed to a warm climate with constant sunlight that favoured biological degradation and photolytic breakdown of PAHs (Lobao et al. 2022).

The relative percentages of PAHs showed a very different behaviour between the crude oil samples exposed in seawater and soil throughout the test due to the intrinsic nature of both systems (Figures 25 and 26). Initially (T0) the presence of naphthalene (N_0) and its methylderivatives, phenanthrene (P_0) and methylphenanthrenes (P_1), dibenzothiophene (D_0) and methyldibenzothiophenes (D_1) was observed. The major compounds were dimethylnaphthalenes (N_2) in the aqueous system (Figure 25A) and trimethylnaphthalenes (N_3) in the lithospheric system (Figure 26A) with relative abundances below 30 % in both cases. For the crude oil samples in seawater, between T1 and T4 (Figures 25B-E) there was a total loss of naphthalene and a decrease of methylnaphthalenes (N_1), in addition, dimethylnaphthalenes (N_2) and dimethylnaphthalenes (N_3) in the lithospheric system (Figure 26A) with relative abundances below 30 %.

(N_2) were exceeded in proportion by trimethylnaphthalenes (N_3). On the other hand, for crude oils extracted from soil, there was a sustained depletion of naphthalene and its methylderivatives that can evaporate from the soil surface or be oxidised by the action of microorganisms (An et al. 2005; López and Infante 2021). This decreasing depletion with the substitution of alkyl groups was associated with an evaporation process that has been observed in other studies carried out in the laboratory (Brakstad et al. 2014, Olson et al. 2017). Also in accidents such as the Deepwater Horizon oil rig, where evaporation of light alkanes and low molecular weight PAHs was concluded to be the main surface oil depletion process (Brown et al. 2011).

Between two and four months (T5 and T6), the histograms (Figures 25F-G) showed first a decrease in methylnaphthalenes and then in dimethylnaphthalenes. A marked decrease in trimethylnaphthalenes and a significant increase in tetramethylnaphthalenes (N4), in phenanthrene and mainly in methylphenanthrenes to become the dominant PAHs with relative percentages above 40 %. It is interesting to note that in another investigation it was determined that PAHs of low molecular mass decreased significantly in crude oil samples exposed to seawater over the course of 60 and 90 days. Yang et al. (2016) attributed this behaviour to the volatilisation of naphthalenes, but not to the rest of the PAHs. In the case of samples contained in reactors with soil, a reduction in the values of naphthalene and its methylated derivatives and, in addition, an increase in phenanthrene and methylphenanthrenes were observed (Figures 26F-G). The adsorption of PAHs to the soil slows down their volatilisation, an effect that increases when there is an abundance of organic matter in the sediment matrix. The Patagonian soil used in this experiment is characterised by a low organic matter content that confers low sorption capacity for organic pollutants (Toledo et al. 2022). In addition, the hydrophobicity of naphthalene, phenanthrene and their methylcompounds determines that their capacity to adsorb on soil particles is slow, therefore, their molecular masses are the main aspect that conditions their behaviour in the face of evaporation (An et al. 2005). Based on the above, the results exhibit an expected behaviour as heavier PAHs remain in the soil longer due to a more stable and hydrophobic molecular structure (Abdulazeez and Fantke 2017).

After T7 and T8 (six and nine months) the histograms showed a decrease of tri- and tetramethylnaphthalenes. In addition, a marked stability of phenanthrene and methylphenanthrene (P1) as the most abundant with percentages around 60 % for the latter group in the crude oil and seawater systems (Figures 25H-I). On the other hand, the soil experience showed some differences as naphthalene and its methylmolecules still remained in the system, but decreasing as a function of time. In addition, there was a steady increase of phenanthrene and methylphenanthrenes (P1) with percentages close to 20 and 50 %, respectively (Figures 26H-I). In many cases, the relative proportions of naphthalenes in relation to other PAH series decrease sharply due to their marked volatility (Kao et al. 2015). Finally, after one year (T9) tri- and tetramethylnaphthalenes decreased considerably in the seawater samples, phenanthrene remained almost unchanged and methylphenanthrenes continued their trend with values above 60 %. Notably, dibenzothiophene (D0) remained almost unchanged over the 12 months with minimal relative concentrations and methyldibenzothiophenes (D1) showed a slight increase over the same period (Figure 25J). Regarding the hydrocarbons disseminated in soil, both naphthalene and its methylated derivatives presented very low values in relation to the initial crude oil samples (T0). On the other hand, phenanthrene and methylphenanthrenes continued to increase their relative abundances (Figure 26J). It is interesting to add that a staircase-like pattern from naphthalene to methylphenanthrenes was observed in Figure 26. This behaviour was reported in a study on crude oil spills in Antarctica, in which volatilisation was attributed to have been the main mechanism of hydrocarbon removal in the affected areas (Aislabie et al. 1999; Rodriguez et al. 2018). With respect to sulphur PAHs, it can be mentioned that dibenzothiophene ($D0$) remained practically unchanged over the study time with a low relative abundance and that methyldibenzothiophenes ($D1$) underwent a gradual decrease over the same time span from 3 to 2 %.

The main conclusions based on the results obtained throughout this thesis are detailed below:

• From the dendogram it can be said that all the crude samples were separated by levels from highest to lowest degree of relatedness according to the following criteria: oil reservoir < depth of puncture < geological formation < palaeobiodegradation.

• The distribution of tetracyclic biomarkers in the 217 ion fragmentogram revealed differences in the samples studied. At the formation level, the Springhill crudes were characterised by the preponderance of pregnanes and the presence of diasteranes and steranes. The Lower Magellan associated hydrocarbons showed a very low concentration of these molecules.

• The results obtained from the characterisation of the biomarker pattern and their diagnostic relationships in the samples studied suggest that all the crude oils studied were generated from a mixed type kerogen. This was characterised by marine organic matter and a secondary input from the mainland, deposited under low oxygenation conditions. The Palermo Aike and Margas Verdes formations, defined as the source rocks for the crude oils present in Springhill and Lower Magallanes respectively, were characterised by a siliciclastic marine nature. The AN crude oil was the only one that showed alterations associated with a paleobiodegradation process.

• During the laboratory-scale weathering test of crude oil samples in seawater and soil, two processes coexisted and altered the composition of the hydrocarbons. In the first instance evaporation has promoted a decrease in light n-alkanes, the observed changes in the RDs of the analysed compounds and the distribution of PAHs in the histograms throughout the simulation. The other process that generated changes was the biodegradation associated with a baseline uplift in TICs that led to the formation of MCNRs. However, most of the biomarkers remained unchanged under the test conditions throughout the study time.

• If we consider that these conditions at the time of the study can be extrapolated to real situations involving oil spills in seawater or soil. Then the potential use of these biomarkers is adequate to solve the aforementioned problems since they constitute a unique and reliable chemical signature known as a fingerprint that identifies each crude oil.

6. BIBLIOGRAPHY

1. Abdulazeez, T., Fantke, P. (2017). Polycyclic aromatic hydrocarbons. A review. Cogent Environmental Science, 3(1).

2. Adedosu, T., Sonibare, O., Tuo, J., Ekundayo, O. (2012). Biomarkers, carbon isotopic composition and source rock potentials of Awgu coals, middle Benue trough, Nigeria. Journal of African Earth Sciences, 66 - 67, 13 - 21.

3. Aeppli, C., Carmichael, C., Nelson, R., Lemkau, K., Graham, W., Redmond, M., Valentine, D., Reddy, C. (2012). Oil Weathering after the Deepwater Horizon Disaster Led to the Formation of Oxygenated Residues. Environmental Science & Technology, 46(16), 8799 - 8807.

4. Agüero-Manzano, Y. (2019). Hydrocarbon-specific compound isotope ratios (CSIA) applied in environmental forensics associated with marine oil spills. Master's thesis. Engineering Area, Universidad Católica Andrés Bello. Caracas, Venezuela, 158 p.

5. Aislabie, J., Balks, M., Astori, N., Stevenson, G., Symons, R. (1999). Polycyclic aromatic hydrocarbons in fuel-oil contaminated soils. Antarctica, 39(13), 0 - 2207.

6. Alberdi, M., Moldowan, J., Peters, K., Dahl J.E. (2001). Stereoselective biodegradation of tricyclic terpanes in heavy oils from the Bolivar Coastal Fields, Venezuela. Organic Geochemistry, 32, 181 - 191.

7. Ali, N., Dashti, N., Khanafer, M., Al-Awadhi, H., Radan, S. (2020). Bioremediation of soils saturated with spilled crude oil. Science Report, 10, 1116.

8. An, T., Chen, H., Zhan, H., Zhu, Z., Berndtsson, R. (2005). Sorption kinetics of naphthalene and phenanthrene in loess soils. Environmental Geochemistry, 47(4), 467 - 474.

9. API - American Petroleum Iinstitute (2016). Sunken Oil Detection and Recovery Operational Guide. Technical Report 1154 - 1 (126 p).

10. Aramendia, I., Ramos, M., Geuna, S., Cuitino, J., Ghiglione, M. (2018). A multidisciplinary study of the Lower Cretaceous marine to continental transition in the northern Austral-Magallanes basin and its geodynamic significance. Journal of South American Earth Sciences, 86, 54 - 69.

11. Arekhi, M., Terry, L., John, G., Prabhakar Clement, T. (2021). Environmental fate of petroleum biomarkers in Deepwater Horizon oil spill residues over the past 10 years. Science of The Total Environment, 148056.

12. Barberón, V., Ronda, G., Leal, P., Sue, C., Ghiglione, M. (2015). Lower Cretaceous provenance in the northern Austral basin of Patagonia from sedimentary petrography. Journal of South American Earth Sciences, 64(2), 498 - 510.

13. Belotti, H., Rodriguez, J., Conforto, G. (2014). The Palermo Aike Formation as an unconventional reservoir in the Austral Basin, Santa Cruz Province, Argentina. IX Hydrocarbon Exploration and Development Congress, Mendoza - Argentina.

14. Bost, F., Frontera-Suau, R., McDonald, J., Peters, K., Morris, P. (2001). Aerobic biodegradation of hopanes and norhopanes in Venezuelan crude oil. Organic Geochemistry 37, 105 - 114.

15. Brakstad, O., Daling, P., Faksness, L., Almas, I., Vang, S., Syslak, L., Leirvik, F. (2014). Depletion and biodegradation of hydrocarbons in dispersions and emulsions of the Macondo 252 oil generated in an oil-on-seawater mesocosm flume basin. Marine Pollution Bulletin, 84(1 - 2), 125 - 134.

16. Brown, J., Beckmann, D., Bruce, L., Cook, L., Mudge, S. (2011). PAH depletion ratios document the rapid weathering and attenuation of PAHs in oil samples collected after the Deepwater Horizon. In: Proceedings of the 2011 International Oil Spill Conference. Portland, Oregon.

17. Cai, M., Yao, J., Yang, H., Wang, R., Masakorala, K. (2013). Aerobic Biodegradation Process of Petroleum and Pathway of Main Compounds in Water Flooding Well of Dagang Oil Field. Bioresource Technology, 144, 100 - 106.

18. Cagnolatti, M., Curia, D. (1990). Transgressive sequences of the Springhill F. in southeastern Santa Cruz Province, Southern Basin, Argentina. III Reunión Argentina de Sedimentología, San Juan,

Argentina, Abstracts: 72 - 80.

19. Cagnolatti, M., Martins, R., Villar, H. (1996). La Formación Lemaire como probable generadora de hidrocarburos en el área Angostura, Provincia de Tierra del Fuego, Argentina, in Décimo Tercer Congreso Geológico Argentino y Tercer Congreso de Exploración de Hidrocarburos: Mar del Plata, Buenos Aires, Argentina, Instituto Argentino del Petróleo y del Gas, 123 - 139.

20. Cagnolatti, M., Miller, M. (2002). Reservoirs of the Magallanes Formation, in Schiuma, M., Hinterwimmer, G., Vergani, G. eds., Reservoir Rocks of the Productive Basins of Argentina. Symposium of the V Hydrocarbon Exploration and Development Congress, 91 - 114.

21. Calderón, M., Hervé, F., Fuentes, F., Fosdick, J., Sepúlveda,F., and Galaz, G. (2016). Tectonic evolution of Paleozoic and Mesozoic andean metamorphic complexes and the Rocas Verdes ophiolites in southern Patagonia. In: Ghiglione, M.C. (Ed.), Geodynamic Evolution of the Southernmost Andes. Springer, Cham, 7 - 36.

22. CEN - CENTER FOR EUROPEAN NORMS (2012). Oil spill identification. Waterborne petroleum and petroleum products. Analytical methodology and interpretation of results based on GC-FID and GC-MS low resolution analyses. CEN Technical Report 15522-2. Brussels, Belgium.

23. Chen, Z., Wang, T., Li, M., Yang, F., Cheng, B. (2018). Biomarker geochemistry of crude oils and Lower Paleozoic source rocks in the Tarim Basin, western China: An oil-source rock correlation study. Marine and Petroleum Geology, 96, 94 - 112.

24. Choi, B., Lee, S., Jho, E. (2020). Removal of TPH, UCM, PAHs, and Alk-PAHs in oil-contaminated soil by thermal desorption. Applied Biological Chemistry, 63(1), 1 - 6.

25. Cortes, J., Rincon, J., Jaramillo, J., Philp, R., Allen, J. (2010). Biomarkers and compound-specific stable carbon isotope of n-alkanes in crude oils from Eastern Llanos Basin, Colombia. Journal of South American Earth Sciences, 29, 198 - 213.

26. Costa de Sousa, A., da Sousa, E., Rodrigues de Sousa, G., Carbonezi, C., Durante, A., Silva, B., de Lima. S. (2022). An alternative method for the separation and analysis of acidic biomarkers from crude oil samples. Journal of South American Earth Sciences, 120, 104054.

27. Cuitiño, J., Varela, A., Ghiglione, M., Richiano, S., Poiré, D. (2019). The Austral- Magallanes Basin (southern Patagonia): asynthesis of its stratigraphy and evolution. Latin American Journal of Sedimentology and Basin Analysis, 26(2), 155 - 166.

28. Cunningham, W., Mann, P. (2007). Tectonics of strike-slip restraining and releasing bends. Geological Society, London, Special Publications, 290(1), 1 - 12.

29. Diraison, M., Cobbold, P., Gapais, D., Rossello, E., Le Corre, C. (2000). Cenozoic crustal thickening, wrenching and rifting in the foothills of the southernmost Andes. Tectonophysics, 316: 91 - 119.

30. Didyk, B., Simoneit, B., Brassell, S., Eglinton, G. (1978). Organic geochemical indicators of palaeoenvironmental conditions of sedimentation. Nature, 272, 216 - 222.

31. Douglas, G., Bence, A., Prince, E., Roger, C., Mc Millen, S., Butler, E. (1996). Environmental Stability of Selected Petroleum Hydrocarbon Source and Weathering Ratios. Environmental Science & Technology, 30(7), 2332 - 2339.

32. Douglas, G., Hardenstine, J., Liu, B., Uhler, A. (2012). Laboratory and field verification of a method to estimate the extent of petroleum biodegradation in soil. Environmental Science & Technology, 46(15), 8279 - 8287.

33. Dow, W. (1977). Kerogen studies and geological interpretations. Journal of Geochemical Exploration, 7, 79 - 99.

34. Dowey, P., Osborne, M., Volk, H. (2020). About this title - Application of Analytical Techniques to Petroleum Systems. Geological Society, London, Special Publications, 484(1).

35. El-Sabagh, S., El-Naggar, A., El Nady, M., Ebiad, M., Rashad, A., Abdullah, E. (2018). Distribution of triterpanes and steranes biomarkers as indication of organic matters input and depositional environments of crude oils of oilfields in Gulf of Suez, Egypt. Egyptian Journal of Petroleum, 27, 969 - 977.

36. Escobar, M., Márquez, G., Azuaje, V., Da silva A., Tocco, R. (2012). Use of biomarkers, porphyrins, and trace elements to assess the origin, maturity, biodegradation, and migration of Alturitas oils in Venezuela. Fuel, 97, 186 - 196.

37. Faboya, O., Sojinu, S., Sonibare, O., Falodun, O., Liao, Z. (2016). Aliphatic biomarkers distribution in crude oil-impacted soils: An environmental pollution indicator. Environmental Forensics, 17(1), 27 - 35.

38. Fang, R., Littke, R., Zieger, L., Baniasad, A., Li, M., Schwarzbauer, J. (2019). Changes of composition and content of tricyclic terpane, hopane, sterane, and aromatic biomarkers throughout the oil window: A detailed study on maturity parameters of Lower Toarcian Posidonia Shale of the Hils Syncline, NW Germany. Organic Geochemistry, 138.

39. Fernández-Varela, R., Andrade, J., Muniategui, S., Prada, D. (2010). Selecting a reduced suite of diagnostic ratios calculated between petroleum biomarkers and polycyclic aromatic hydrocarbons to characterize a set of crude oils. Journal of Chromatography A, 1217, 8279 -8289.

40. Figari, E., Strelkov, E., Cid de la Paz, M., Laffitte, G., Villar, H. (2002). Golfo San Jorge Basin: structural, stratigraphic and geochemical synthesis. Geology and Natural Resources of Santa Cruz. Report of the XV Argentine Geological Congress (Ed: Haller, M.). El Calafate, AGA, 571 - 601p.

41. Gaines, S., Eglinton, G., Rullkötter, J. (2009). Echoes of life: what fossil molecules reveal about earth history. United Kingdom: Oxford University Press.

42. Gallardo, R. (2014). Seismic sequence stratigraphy of a foreland unit in the Magallanes- Austral Basin, Dorado Riquelme Block, Chile: implications for deepmarine reservoirs. Latin American Journal of Sedimentology and Basin Analysis, 21(1), 49 - 64.

43. Garcette-Lepecq, A., Derenne, S., Largeau, C., Bouloubassi, I., Saliot, A. (2000). Origin and formation pathways of kerogen like organic matter in recent sediments off the Danube Delta (northwestern Black Sea). Organic Geochemistry, 31, 1663 - 1683.

44. Garcia, M., Cattani, A., da Cunha Lana, P., Figueira, R., Martins, C. (2019). Petroleum biomarkers as tracers of low-level chronic oil contamination of coastal environments: A systematic approach in a subtropical mangrove. Environmental Pollution, 249, 1060 - 1070.

45. Garrett, R., Pickering, I., Haith, C., Prince, R. (1998). Photooxidation of crude oils. Environmental Science & Technology, 32(23), 3719 - 3723.

46. Giacosa, R., Fracchia, D., Heredia, N. (2012). Structure of the Southern Patagonian Andes at 49°S. Geologic Acta, 10: 265 - 282.

47. Greenwood, P., Wibrow, S., Suman, J., Tibbett, M. (2008). Sequential hydrocarbon biodegradation in a soil from arid coastal Australia treated with oil under laboratory- controlled conditions. Organic Geochemistry, 39(9), 0 - 1346.

48. Han, Y., Nambi, I., Clement, T. (2018). Environmental impacts of the Chennai oil spill accident-a case study. Science of The Total Environmental, 626, 795 - 806.

49. Han, Y., John, G., Clement, T. (2019). Understanding the thermal degradation patterns of hopane biomarker compounds present in crude oil. Science of The Total Environment, 667, 792 - 798.

50. Hardebol, N., Callot, J., Bertotti, G., Faure, J. (2009). Burial and temperature evolution in thrust belt systems: Sedimentary and thrust sheet loading in the SE Canadian Cordillera. Tectonics, 28(3).

51. He, D., Simoneit, B., Cloutier, J., Jaffé, R. (2018). Early diagenesis of triterpenoids derived from mangroves in a subtropical estuary. Organic Geochemistry, 125, 196e211.

52. Hughes, W., Holba, A., Dzou, L. (1995). The ratios of dibenzothiophene to phenanthrene and pristane to phytane as indicators of depositional environment and lithology of petroleum source rocks. Geochimica et Cosmochimica Acta, 59, 3581 - 3598.

53. Inglis, G., Naafs, B., Zheng, Y., McClymont, E., Evershed, R., Pancost, R. (2018). Distributions of geohopanoids in peat: implications for the use of hopanoid-based proxies in natural archives. Geochimica et Cosmochimica Acta, 224, 249e261.

54. Jarvie, D., Hill, R., Ruble, T., Pollastro, R. (2007). Unconventional shale-gas systems: The Mississippian Barnett Shale of north-central Texas as one model for thermogenic shale-gas

assessment. AAPG Bulletin, 91(4), 475 - 499.

55. John, G., Han, Y., Clement, T. (2018). Fate of hopane biomarkers during in-situ burning of crude oil-a laboratory-scale study. Marine Pollution Bulletin, 133, 756 - 761.

56. Joo, C., Shim, W., Kim, G., Ha, S., Kim, M., An, J., Kim, E., Kim, B., Jung, S., Kim, Y., Yim, U. (2013). Mesocosm Study on Weathering Characteristics of Iranian Heavy Crude Oil with and without Dispersants. Journal of Hazardous Materials, 248, 37 - 46.

57. Kao, N., Su, M., Chang, C., Yen, C. (2018) Revealingminor terpane biomarkers in lubricants and soils using the customized cleanup method.J Soils Sediments 18, 136 - 147.

58. Kao, N., Su, M., Fan, J., Chung, Y. (2015). Identification and quantification of biomarkers and polycyclic aromatic hydrocarbons (PAHs) in an aged mixed contaminated site: from source to soil. Environmental Science and Pollution Research, 22(10), 7529 - 7546.

59. Kienhuis, P., Kraus, U., Kooistra, K. (2019). Oil identification. In Worsfold, P., Townshend, A., Poole, C., Miro, M. (Eds.). Encyclopedia of Analytical Science. United Kingdom: Academic Press.

60. Killops, S., Killops, V. (2005). Introduction to Organic Geochemistry. United Kingdom: Blackwell Publishing.

61. Kim, S., Stanford, L., Rodgers, R., Marshall, A., Walters, C., Kuangnan, Q., Wenger, L., Mankiewicz, P. (2005). Microbial alteration of the acidic and neutral polar NSO compounds revealed by Fourier transform ion cyclotron resonance mass spectrometry. Organic Geochemistry, 36, 1117 - 1134.

62. Khan, M., Biswas, B., Smith, E., Naidu, R., Megharaj, M. (2018). Toxicity assessment of fresh and weathered petroleum hydrocarbons in contaminated soil - a review. Chemosphere, 212, 755 - 767.

63. Klemt, W., Kay, M., Wiklund, J., Wolfe, B., Hall, R., (2020). Assessment of vanadium and nickel enrichment in lower Athabasca River floodplain lake sediment within the athabasca oil sands region (Canada). Environmental Pollution, 265.

64. Kujawinski, E., Freitas, M., Zang, X., Hatcher, P. (2002). The application of electrospray mass spectrometry (ESI MS) to the structural characterization of natural organic matter. Organic Geochemistry, 33, 171 - 180.

65. Kuppusamy, S., Maddela, N. R., Megharaj, M., & Venkateswarlu, K. (2020). Total petroleum hydrocarbons: environmental fate, toxicity, and remediation. Springer.

66. Legarreta, L., Villar, H. (2011). Geological and Geochemical Keys of the Potential Shale Resources, Argentina Basins. Unconventional Resources, Basics, Challenges, and Opportunities for New Frontier Plays, Buenos Aires - Argentina.

67. Lemkau, K., Peacock, E., Nelson, R., Ventura, G., Kovecses, J., Reddy, C. (2010). The M/V Cosco Busan spill: Source identification and short-term fate. Marine Pollution Bulletin 60(11), 2123 - 2129.

68. Li, Bocai, He, Daxiang, Li, Meijun, Chen, Lin, Yan, Kai, Tang, Youjun (2022). Biomarkers and Carbon Isotope of Monomer Hydrocarbon in Application for Oil - Source Correlation and Migration in the Moxizhuang - Yongjin Block, Junggar Basin, NW China. ACS Omega, 7(50), 47317 - 47329.

69. Lobao, M., Thomazelli, F., Batista, E., Oliveira, R., Souza, M., Matos, N. (2022). Chronic oil spills revealed by the most important set of samples from the incident in northeastern Brazil, 2019. Anais da Academia Brasileira de Ciencias, 94, e20210492.

70. López, L., Lo Mónaco, S. (2010). Geochemistry of crude oils from the Orinoco Oil Belt, Eastern Basin of Venezuela. Revista de la Facultad de Ingeniera Universidad Central de Venezuela, 25(2), 41 - 50.

71. López, L., Lo Mónaco, S. (2017). Vanadium, nickel and sulfur in crude oils and source rocks and their relationship with biomarkers: Implications for the origin of crude oilsin Venezuelan basins. Organic Geochemistry, 104, 53 - 68.

72. López, L., Infante C. (2021). Changes in biomarkers of the saturated hydrocarbon fraction in a bioremediation test with an extra-heavy crude oil. International Journal of Environmental Pollution, 37, 119 - 131.

73. López, L., Lo Mónaco, S., Kalkreuth, W., Peralba, M. (2019). Assessment of the depositional environment and source rock potential of Permian shales, siltstones and coal seams from the Santa Terezinha Coalfield, Paraná Basin, Brazil. Journal of South American Earth Sciences, 94, 102227.

74. Lorenzo, E., Roca-Beltrán, W., Martínez, M., Morato, A., Escandón-Panchana, P., Álvarez-Domínguez, C. (2018). Geochemical correlation between crude oils and rocks of the petroleum system of the Santa Elena Peninsula and the Gulf of Guayaquil. Geology Bulletin, 40(1), 31 - 42.

75. Lundberg, R. (2019). Validation of Biomarkers for the Revision of the CEN/TR 15522-2:2012. Method: A Statistical Study of Sampling, Discriminating Powers and Weathering of new Biomarkers for Comparative Analysis of Lighter Oils. Student thesis, Linkoping University, Sweden.

76. Malmborg, J., Kooistra, K., Kraus, U., Kienhuis, P. (2020). Evaluation of light petroleum biomarkers for the 3rd edition of the European Committee for Standardization methodology for oil spill identification (EN15522-2). Environmental Forensics, 22(3 - 4), 325 - 339.

77. Mariano, A., Kataoka, A., Angelis, D., Bonotto, D. (2007). Laboratory study on the bioremediation of diesel oil contaminated soil from a petrol station. Brazilian Journal of Microbiology, 38, 346 - 353.

78. Moldowan, J., Seifert, W., Gallegos, E. (1985). Relationship between petroleum composition and depositional environment of petroleum source rocks. AAPG Bulletin, 69(8), 1255 - 1268.

79. Mpodozis, C., Mella, P., Padva, D. (2011). Stratigraphy and sedimentary megasequences in the Austral - Magallanes Basin, Argentina and Chile, in Séptimo Congreso de Exploración y Desarrollo de Hidrocarburos: Mar del Plata, Buenos Aires, Argentina, Instituto Argentino del Petróleo y del Gas, 97 - 137.

80. Oliveira, O., Queiroz, A., Cerqueira, J., Soares, S., Garcia, K., Pavani Filho, A., Rosa, M., Suzart, C., Pinheiro, L., Moreira, I. (2020). Environmental disaster in the northeast coast of Brazil: Forensic geochemistry in the identification of the source of the oily material. Marine Pollution Bulletin, 160(111597), 1 - 7.

81. Olson, G., Gao, H., Meyer, B., Miles, M., Overton, E. (2017). Effect of Corexit 9500 A on Mississippi Canyon crude oil weathering patterns using artificial and natural seawater. Heliyon, 3(3), e00269.

82. Orea, M., López, L., Ranaudo, M., Faraco, A. (2021). Saturated biomarkers adsorbed and occluded by the asphaltenes of some Venezuelan crude oils: Limitations in geochemical assessment and interpretations. Journal of Petroleum Science and Engineering, 206, 109048.

83. Orta-Martínez, M., Rosell-Melé, A., Cartró-Sabaté, M., O'Callaghan-Gordo, C., Moraleda-Cibrián, N., Mayor, P. (2018). First evidences of Amazonian wildlife feeding on petroleum-contaminated soils: a new exposure route to petrogenic compounds? Environmental Research, 160, 514 - 517.

84. Peters, K., Walters, C., Moldowan, J. (2005). The Biomarker Guide. Biomarkers and Isotopes in the Environment and Human History. United Kingdom: Cambridge University Press.

85. Pittion, J., Goudaian, J. (1992). Source-rocks and oil generation in the Austral basin. XIII Word Petroleum Congress, Buenos Aires, 2, 113 - 120.

86. Pittion, J., Arbe, H., (1999). Petroleum system of the Austral Basin. IV Hydrocarbon Exploration and Development Congress, IAPG, Proceedings I, 239 - 250.

87. Poiré, D., Franzese, J. (2010). Mesozoic clastic sequences from a Jurassic rift to Cretaceous foreland basin, Austral Basin, Patagonia, Argentina. In: del Papa, C. and R. Astini (Eds), Field Excursion Guidebook, 18th International Sedimentological Congress, Argentina. FE-C13, 53 pp.

88. Prince, R., Elmendorf, D., Lute, J., Hsu, C., Haith, C., Senius, J., Dechert, G., Douglas, G., Butler, E. (1994). 17-α(H)- 21-β(H)-Hopane as a conserved internal marker for estimating the biodegradation of crude oil. Environmental Science & Technology 28(1), 142 - 145.

89. Ramos, M., Suárez, R., Boixart, G., Ghiglione, M., Ramos, V. (2019). The structure of the northern Austral Basin: Tectonic inversion of Mesozoic normal faults. Journal of South American Earth Sciences, 94.

90. Rangel, A., Osorno, J., Ramirez, J., De Bedout, J., González, J., Pabón, J. (2017). Geochemical assessment of the Colombian oils based on bulk petroleum properties and biomarker parameters. Marine and Petroleum Geology, 86, 1291 - 1309.

91. Reyes, C., Moreira, I., Oliveira, D., Medeiros, N., Almeida, M., Wandega, F., Soares, S., Oliveira, O. (2014). Weathering of Petroleum Biomarkers: Review in Tropical Marine Environment Impacts. Open Access Library Journal, 1, 1 - 2.

92. Richardson, J., Miller, D. (1982). Identification of Dicyclic and Tricyclic Hydrocarbons in the Saturate Fraction of a Crude Oil by Gas Chromatography/Mass. Analytical Chemistry, 54, 765 - 768.

93. Richiano, S., Varela, A., Cereceda, A., Poiré, D. (2012). Paleoenvironmental evolution of the Río Mayer Formation, Lower Cretaceous, Southern Basin, Santa Cruz Province, Argentina. Latin American Journal of Sedimentology and Basin Analysis, 19(1), 3 - 26.

94. Richiano, S., Varela, A., Poiré, D. (2016). Heterogeneous distribution of trace fossils across initial transgressive deposits in rift basins: an example from the Springhill Formation, Argentina. Lethaia, 49(4), 524 - 539.

95. Robbiano, J., Arbe, H., Gangui, A. (1996). Cuenca Austral Marina, in Ramos, V. y Turic., M., eds., Geologla y Recursos Naturales de Ia Plataforma Continental Argentina, Relatorio del XIII Congreso Geológico Argentino y III Congreso de Exploración de Hidrocarburos, Buenos Aires, Argentina, 343 - 358.

96. Rodríguez, J., Miller, M., Cagnolatti, M. (2008). Petroleum Systems of the Austral Basin, Argentina and Chile. In C. E. Cruz, J. F. Rodríguez, J. J. Hechem and H. J. Villar, eds., Symposium: "Petroleum Systems of the Andean Basins". Instituto Argentino del Petróleo y del Gas. Talleres Trama S.A., Buenos Aires.

97. Rodríguez, C., Iglesias, K., Bícego, M., Taniguchi, S., Sasaki, S., Kandratavicius, N., Venturini, N. (2018). Hydrocarbons in soil and meltwater stream sediments near Artigas Antarctic Research Station: origin, sources and levels. Antarctic Science, 30, 170 - 182.

98. Rodríguez, R., Hernández, A., Domínguez, Z., Nuñez Clemente, A., del Sol Ortega, O. (2020). Analytical methodology for the analysis of terpanes and steranes biomarkers in crude oil samples by gas chromatography coupled to mass spectrometry. Revista CENIC Ciencias Químicas, 51(2), 209 - 223.

99. Ron, E., Rosenberg, E. (2014). Enhanced Bioremediation of Oil Spills in the Sea. Current Opinion in Biotechnology, 27, 191-194.

100. Rosell-Melé, A., Moraleda-Cibrián, N., Cartró-Sabaté, M., Colomer-Ventura, F., Mayor, P., Orta-Martínez, M. (2018). Oil pollution in soils and sediments from the Northern Peruvian Amazon. Science of The Total Environment, 1010 - 1019.

101. Rullkotter, J., Welte, D. (1980). Oil-oil and oilcondensate correlation by low eV GC- MS measurements of aromatic hydrocarbons, in Douglas, A., Maxwell, J. (eds). Advances in Organic Geochemistry (1979). Physics and Chemistry of the Earth, 12, 93 - 102.

102. Salat, A., Eickmeyer, D., Kimpe, L., Hall, R., Wolfe, B., Mundy, L., Blais, J. (2020). Integrated analyses of petroleum biomarkers and polycyclic aromatic compounds in lake sediment cores from an oil sands region. Environmental Pollution, 116060.

103. Schito, A., Corrado, S. (2018). An automatic approach for characterization of the thermal maturity of dispersed organic matter Raman spectra at low diagenetic stages. Geological Society, London, Special Publications, SP484.5.

104. Schiuma, M., Hinterwimmer, G., Vergani, G. (2018). Reservoir Rocks of the Productive Basins of Argentina: Buenos Aires, Instituto Argentino del Petróleo y del Gas, 1006 p.

105. Schwarzkopf, T., Leythaeuser, D. (1988). Oil generation and migration in the Gifhorn Trough, NW-Germany: Organic Geochemistry 13(1 - 3), 245 - 253.

106. Seguel, C., Mudge, S., Salgado, C., Toledo, M. (2001). Tracing sewage in the marine environment: Altered signatures in Concepcion Bay, Chile. Water Research, 17, 4166 - 4174.

107. Seifert, W., Moldowan, J. (1978). Aplication of steranes, triterpane and monoaromatics to the

maturation of crude oils. Geochimica et Cosmochimica Acta, 42, 71 - 95.

108. Seifert, W., Moldowan, J., Demaison, G. (1984). Source correlation of biodegraded oils. Organic Geochemistry, 6, 633 - 643.

109. Shanmugam, G. (1985). Significance of coniferous rain forests and related organic matter in generating commercial quantities of oil, Gippsland Basin, Australia. American Association of Petroleum Geologists Bulletin, 69, 1241 - 1254.

110. Sivan, P., Datta, G., Singh, R. (2008). Aromatic biomarkers as indicators of source, depositional environment, maturity, and secondary migration in the oils of Cambay Basin, India. Organic Geochemistry, 39(11), 1620 - 1630.

111. Song, X., Zhang, B., Chen, B. Cai, Q. (2016). Use of sesquiterpanes, steranes, and terpanes for forensic fingerprinting of chemically dispersed oil. Water, Air and Soil Pollution, 227(8), 281.

112. Speight, J. (2014). The chemistry and technology of petroleum. CRC press.

113. Stashenko, E., Martínez, J., Robles, M. (2014). Selective extraction and specific detection of saturated biomarkers from oil. Scientia Chromatographica 6, 251 - 268.

114. Stout, S., Wang, Z. (2007). Chemical fingerprinting of spilled or discharged petroleum - methods and factors affecting petroleum fingerprints in the environment. In: Wang, Z., Stout, S., (Eds). Oil Spill Environmental Forensics. United Kingdom: Academic Press.

115. Stout, S., Wang, Z. (2016). Chemical fingerprinting methods and factors affecting petroleum fingerprints in the environment. In: Stout, S., Wang, Z. (Eds). Standard Handbook Oil Spill Environmental Forensics Fingerprinting and Source Identification. United Kingdom: Academic Press.

116. Summons, R., Lincoln, S. (2012). Biomarkers: informative molecules for studies in geobiology, in Knoll, A., Canfield, D., Konhauser, K. (eds). Fundamentals of Geobiology: Oxford, Wiley-Blackwell, 269-298.

117. Sutton, P., Lewis, C., Rowland, S. (2005). Isolation of individual hydrocarbons from the unresolved complex hydrocarbon mixture of a biodegraded crude oil using preparative capillary gas chromatography. Organic Geochemistry, 36, 963 - 970.

118. Swannel, R., Lee, K., McDonagh, M. (1996). Field evaluation of marine oil spill bioremediation. Microbiological Reviews, 60, 342 - 365.

119. Schwarz, E., Veiga, G., Spalletti, L., Massaferro, J. (2011). The transgressive infill of an inherited-valley system: The Springhill Formation (lower Cretaceous) in southern Austral Basin, Argentina. Marine and Petroleum Geology, 28: 1218 - 1241.

120. Tissot, B., Welte, D. (1984). Petroleum formation and occurrence. Berlin, SpringerVerlag, 666 p.

121. TNRCC (2000). TNRCC METHOD 1006. Characterization of Nc6 to Nc35 Petroleum Hydrocarbons in Environmental Samples. Texas Natural Resource Conservation Commission. Method. United States, Texas, 21 p.

122. Toledo, S., Peri, P., Fontenla, S. (2022). Environmental Conditions and Grazing Exerted Effects on Arbuscular Mycorrhizal in Plants at Southern Patagonia Rangelands. Rangeland Ecology & Management, 81, 44 - 54.

123. Tomas, G., Vargas, W., Acuña, A. (2020). Biomarker geochemical evaluation of the Mosquito deposit in the Southern Basin of Patagonia Argentina: Revista de la Sociedad Geológica de España, 33(2), 31 - 40.

124. Tomas, G., Acuña, A. (2023). Study of oil biomarkers from the weathering of a crude oil in seawater. International Journal of Environmental Pollution, 39, 71 - 84.

125. Truskewycz, A., Gundry, T., Khudur, L., Kolobaric, A., Taha, M., Aburto-Medina, A., Ball, A., Shahsavari, E. (2019). Petroleum hydrocarbon contamination in terrestrial ecosystems - fate and microbial responses. Molecules, 24(18), 3400.

126. Turner, R., Overton, E., Meyer, B., Miles, M., Hooper-Bui, L., Engel, A., Swenson, E., Lee, J., Milan C., Gao, H. (2014). Distribution and recovery trajectory of Macondo (Mississippi Canyon 252) oil in Louisiana Coastal Wetlands. Marine Pollution Bulletin, 87(1 - 2), 57 - 67.

127. U.S. EPA (2007). "Method 3545A (SW-846): Pressurized Fluid Extraction (PFE)," Revision 1.

Washington, DC.

128. Vandenbroucke, M., Largeau, C. (2007). Kerogen origin, evolution and structure. Organic Geochemistry, 38(5), 719 - 833.

129. Vargas-Escudero, M., Ríos-Reyes, C., García-González, M., Ortiz-Orduz, A. (2021). Diagenesis and thermal maturity of the Cogollo Group sedimentary rocks in the ANH- CR-Montecarlo-1X well, Cesar-Ranchería Basin, Colombia. Andean Geology 48(3), 472 - 495.

130. Venosa, A., Suidan, M., King, D., Wrenn, B. (1997). Use of hopane as a conservative biomarker to monitor the effectiveness of bioremediation of crude oil contaminating a sandy beach. Journal of Industrial Microbiology and Biotechnology 18(2 - 3), 131 - 139.

131. Villar, H., Arbe, H. (1993). Oil generation in the Esperanza area, Austral Basin, Argentina (abstract), in Mello M., Trindade L. (eds). Third Latin American Congress on Organic Geochemistry (1992), Manaus, Brazil, 150 - 153.

132. Villar, H., Sylwan, C., Gutiérrez Pleimling, A., Miller, M., Castaño, J., Dow, W. (1996). Formation of heavy oils from biodegradation and mixing processes in the D-129-Cañadón Seco oil system, South Flank of the Golfo San Jorge Basin, Santa Cruz Province, Argentina. Proceedings I, XIII Argentine Geological Congress and III Hydrocarbon Exploration Congress, AGA, IAPG, 45 - 60 p.

133. Walters, C., Wang, F., Higgins, M., Madincea, M. (2018). Universal Biomarker Analysis: Aromatic hydrocarbons. Organic Geochemistry, 124, 205 - 214.

134. Wang, Z., Fingas, M., Blenkinsopp, S., Sergy, G., Landriault, M., Sigouin, L., Foght, J., Semple, K., Westlake, D. (1998). Oil composition changes due to biodegradation and differentiation between these changes to those due to weathering. Journal of Chromatography A, 809, 89 - 107.

135. Wang, Z., Fingas, M., Sigouin, L. (2000). Characterization and source identification of an unknown spilled oil using fingerprinting techniques by GC-MS and GC-FID. Lc-Gc North America, 10, 1058 - 1067.

136. Wang, Z., Fingas, M., Owens, E., Sigouin, L., Brown, C. (2001). Long-term fate and persistence of the spilled Metula oil in a marine salt marsh environment - degradation of petroleum biomarkers. Journal of Chromatography A, 926(2), 275 - 290.

137. Wang, Z., Fingas, M. (2003). Development of oil hydrocarbon fingerprinting and identification techniques. Marine Pollution Bulletin, 47 (9 - 12), 423 - 452.

138. Wang, Z., Stout, S., Fingas, M. (2006). Forensic fingerprinting of biomarkers for oil spill characterization and source identification. Environmental Forensics, 7, 105 - 146.

139. Wang, Z., Yang, C., Yang, Z., Brown, C. (2007). Petroleum biomarker fingerprinting for oil spill characterization and source identification. Oil Spill Environmental Forensics 3, 73 - 146.

140. Yang, Z., Hollebone, B., Brown, C., Yang, C., Wang, Z., Zhang, G., Shah, K. (2016). The photolytic behavior of diluted bitumen in simulated seawater by exposed to the natural sunlight. Fuel, 186, 128 - 139.

141. Yang, Z., Shah, K., Courtemanche, C., Hollebone, B., Yang, C., Beaulac, V. (2023). The fate and behavior of petroleum biomarkers in diluted bitumen and conventional crude oil exposed to natural sunlight in simulated seawater. Chemosphere, 320, 137906.

142. Yavari, S., Malakahmad, A., Sapari, N. (2015). A review on phytoremediation of crude oil spills. Water, Air, and Soil Pollution, 226(8).

143. Yim, U., Ha, S., An, J., Won, J., Han, G., Hong, S., Kim, M., Jung, J., Shim, W. (2011). Fingerprint and weathering characteristics of stranded oils after the Hebei Spirit oil spill. Journal of Hazardous Materials, 197, 60 - 69.

144. Zerfass, H., Ramos, V., Ghiglione, M., Naipauer, M., Belotti, H., Carmo, I. (2017). Folding, thrusting and development of push-up structures during the Miocene tectonic inversion of the Austral basin, southern Patagonian Andes (50 °S). Tectonophysics, 699, 102 - 120.

145. Zhang, H., Yin, X., Zhou, H., Wang, J., Han, L. (2015). Weathering Characteristics of Crude Oils from Dalian Oil Spill Accident, China. Aquatic Procedia, 3, 238 - 244.

7. ANNEX

ANNEX 1

Table AN. Relative standard deviations obtained from the relative abundances of AN crude oil.

RAs	AN1	AN2	AN3	AN4	Averages	SDs	RSDs
n-C9	0,0392	0,0465	0,0343	0,0406	0,0402	0,0050	12,4
n-C10	0,0653	0,0658	0,0633	0,0599	0,0636	0,0027	4,2
n-C11	0,1810	0,1722	0,1450	0,1708	0,1673	0,0155	9,3
n-C12	0,1164	0,1133	0,1139	0,1146	0,1146	0,0013	1,2
n-C13	0,0403	0,0451	0,0492	0,0455	0,0450	0,0037	8,1
n-C14	0,0193	0,0245	0,0217	0,0189	0,0211	0,0026	12,1
n-C15	0,1162	0,1044	0,1007	0,1020	0,1058	0,0071	6,7
n-C16	0,0629	0,0722	0,0696	0,0689	0,0684	0,0039	5,8
n-C17	0,0639	0,0634	0,0652	0,0681	0,0652	0,0021	3,3
P	0,1599	0,1560	0,1907	0,1677	0,1686	0,0155	9,2
n-C18	0,0394	0,0387	0,0375	0,0410	0,0391	0,0015	3,8
F	0,0963	0,0979	0,1090	0,1017	0,1012	0,0057	5,6
n-C19	0,0000	0,0000	0,0000	0,0000	0,0000	0,0000	0,0
n-C20	0,0000	0,0000	0,0000	0,0000	0,0000	0,0000	0,0
n-C21	0,0000	0,0000	0,0000	0,0000	0,0000	0,0000	0,0
n-C22	0,0000	0,0000	0,0000	0,0000	0,0000	0,0000	0,0
n-C23	0,0000	0,0000	0,0000	0,0000	0,0000	0,0000	0,0
n-C24	0,0000	0,0000	0,0000	0,0000	0,0000	0,0000	0,0
n-C25	0,0000	0,0000	0,0000	0,0000	0,0000	0,0000	0,0
n-C26	0,0000	0,0000	0,0000	0,0000	0,0000	0,0000	0,0
n-C27	0,0000	0,0000	0,0000	0,0000	0,0000	0,0000	0,0
n-C28	0,0000	0,0000	0,0000	0,0000	0,0000	0,0000	0,0
n-C29	0,0000	0,0000	0,0000	0,0000	0,0000	0,0000	0,0
n-C30	0,0000	0,0000	0,0000	0,0000	0,0000	0,0000	0,0
T19	0,0171	0,0180	0,0149	0,0156	0,016	0,001	8,6
T20	0,0217	0,0216	0,0247	0,0205	0,022	0,002	8,1
T21	0,0324	0,0360	0,0391	0,0355	0,036	0,003	7,7
T23	0,0388	0,0393	0,0412	0,0397	0,040	0,001	2,7
T24	0,0295	0,0282	0,0260	0,0243	0,027	0,002	8,5
T25	0,0068	0,0081	0,0080	0,0079	0,008	0,001	8,1
T26(R)	0,0387	0,0417	0,0394	0,0435	0,041	0,002	5,4
T26(S)	0,0095	0,0103	0,0091	0,0094	0,010	0,000	5,1
Ts	0,0526	0,0505	0,0505	0,0437	0,049	0,004	7,9
Tm	0,0614	0,0505	0,0607	0,0543	0,057	0,005	9,3
H29	0,2232	0,2242	0,2238	0,2342	0,226	0,005	2,3
H30	0,2893	0,2820	0,2767	0,2740	0,280	0,007	2,4
M30	0,0181	0,0168	0,0165	0,0154	0,017	0,001	6,6
H31(S)	0,0425	0,0476	0,0517	0,0544	0,049	0,005	10,5
H31(R)	0,0359	0,0380	0,0330	0,0361	0,036	0,002	5,8
G30	0,0091	0,0088	0,0086	0,0091	0,009	0,000	2,9
H32(S)	0,0291	0,0312	0,0312	0,0331	0,031	0,002	5,3
H32(R)	0,0230	0,0230	0,0205	0,0268	0,023	0,003	11,1
H33(S)	0,0115	0,0139	0,0149	0,0140	0,014	0,001	10,7
H33(R)	0,0098	0,0103	0,0095	0,0083	0,009	0,001	9,0
S20	0,097	0,079	0,086	0,087	0,087	0,007	8,5
S21	0,077	0,077	0,081	0,070	0,076	0,005	6,0
S22	0,045	0,050	0,051	0,045	0,048	0,003	7,0
D27(βs)	0,120	0,120	0,116	0,106	0,115	0,007	5,7
D27(βr)	0,048	0,048	0,048	0,045	0,047	0,002	3,3
D27(αs)	0,019	0,025	0,024	0,022	0,022	0,003	12,9
D27(ar)	0,019	0,022	0,020	0,018	0,020	0,002	9,4
S27(as)	0,038	0,043	0,044	0,043	0,042	0,002	5,8
S27(βr)	0,099	0,113	0,115	0,112	0,110	0,007	6,3
S27(βs)	0,056	0,066	0,062	0,060	0,061	0,004	6,5
S27(ar)	0,040	0,043	0,043	0,046	0,043	0,003	6,2
S28(as)	0,004	0,004	0,003	0,003	0,004	0,000	8,0
S28(βr)	0,046	0,039	0,040	0,040	0,041	0,003	7,4
S28(βs)	0,023	0,023	0,023	0,024	0,023	0,001	2,3
S28(ar)	0,052	0,045	0,044	0,049	0,047	0,004	7,8
S29(as)	0,084	0,083	0,082	0,093	0,085	0,005	6,0
S29(βr)	0,049	0,038	0,038	0,040	0,041	0,005	12,5
S29(βs)	0,016	0,014	0,014	0,017	0,015	0,002	9,9
S29(ar)	0,068	0,067	0,066	0,080	0,070	0,006	8,8
N	0,006	0,006	0,007	0,006	0,006	0,001	10,8
2-MN	0,023	0,025	0,025	0,023	0,024	0,001	3,6
1-MN	0,023	0,030	0,028	0,025	0,026	0,003	12,4
2-EN	0,005	0,006	0,007	0,006	0,006	0,001	12,4
1-EN	0,009	0,010	0,011	0,009	0,010	0,001	10,6
2.6 + 2.7 - DMN	0,019	0,020	0,024	0,021	0,021	0,002	10,1

1.3 + 1.7 - DMN	0,037	0,035	0,036	0,031	0,035	0,003	8,3
1,6 - DMN	0,014	0,017	0,015	0,014	0,015	0,002	10,1
1,4 + 2,3 - DMN	0,012	0,013	0,014	0,012	0,013	0,001	8,5
1,5 - DMN	0,007	0,008	0,009	0,008	0,008	0,001	7,5
1,2 - DMN	0,017	0,021	0,021	0,018	0,020	0,002	10,1
1,3,7 - TMN	0,081	0,093	0,094	0,082	0,087	0,007	8,0
1,3,6 - TMN	0,020	0,025	0,025	0,022	0,023	0,002	10,7
1,3,5 + 1,4,6 - TMN	0,029	0,035	0,037	0,031	0,033	0,004	11,2
2,3,6 - TMN	0,035	0,040	0,041	0,034	0,038	0,003	9,1
1,2,7 + 1,6,7 - TMN	0,016	0,019	0,020	0,017	0,018	0,002	11,5
1,2,6 - TMN	0,006	0,006	0,006	0,007	0,006	0,000	4,1
1,2,4 - TMN	0,010	0,011	0,013	0,012	0,011	0,001	9,2
1,2,5 - TMN	0,073	0,090	0,088	0,080	0,083	0,008	9,7
Ph	0,105	0,089	0,092	0,100	0,097	0,007	7,7
3-MP	0,102	0,088	0,084	0,096	0,093	0,008	8,6
2-MP	0,117	0,093	0,096	0,115	0,105	0,012	11,8
9-MP	0,117	0,104	0,095	0,106	0,106	0,009	8,6
1-MP	0,061	0,063	0,058	0,071	0,063	0,005	8,5
D	0,001	0,001	0,001	0,001	0,001	0,000	4,9
4-MDBT	0,039	0,035	0,034	0,037	0,036	0,002	5,8
2 + 3 - MDBT	0,011	0,011	0,012	0,012	0,012	0,001	8,3
1 - MDBT	0,005	0,005	0,005	0,004	0,005	0,000	8,9

Idem Table 11.

Table AN. Diagnostic ratios (DRs) obtained from AN crude oil.

RDs	AN1	AN2	AN3	AN4	Prom	SDs	RSDs
P/F	1,661	1,594	1,750	1,649	1,664	0,065	3,9
P/n-C17	2,502	2,462	2,925	2,462	2,588	0,226	8,7
FZn-Cx8	2,446	2,529	2,909	2,480	2,591	0,215	8,3
n-C29/n-C17	0,000	0,000	0,000	0,000	0,000	0,000	0,0
H29/H30	0,771	0,795	0,809	0,855	0,808	0,035	4,4
10 x G3oG3o x C3o	0,063	0,060	0,060	0,056	0,060	0,003	4,3
M3o/H3o	0,306	0,304	0,300	0,321	0,308	0,009	3,0
% S27	40,59	45,85	45,91	43,07	43,86	2,548	5,8
% S28	21,53	19,09	19,25	19,07	19,73	1,197	6,1
% S29	37,87	35,04	34,82	37,85	36,39	1,692	4,6
D27/S27	0,878	0,811	0,790	0,730	0,802	0,061	7,6
IMP	1,158	1,063	1,099	1,141	1,115	0,043	3,8
Rc	1,077	1,025	1,045	1,068	1,053	0,024	2,2
% 4-MeDBT	70,91	69,22	66,15	68,68	68,74	1,971	2,9
% 2+3-MeDBT	19,69	21,72	24,12	23,42	22,24	1,975	8,9
% 1-MeDBT	9,386	9,053	9,721	7,889	9,012	0,797	8,8
DBTZPh	0,013	0,015	0,014	0,015	0,014	0,001	7,9

Idem Table 12.

Table AC. Relative standard deviation obtained from the relative abundances of crude oil AC.

RAs	AC1	AC2	AC3	AC4	Averages	SDs	RSDs
n-C9	0,033	0,033	0,034	0,038	0,034	0,002	6,4
n-C10	0,051	0,048	0,049	0,056	0,051	0,004	7,0
n-C11	0,069	0,060	0,068	0,062	0,064	0,004	6,8
n-C12	0,077	0,071	0,072	0,067	0,072	0,004	5,9
n-C13	0,071	0,072	0,072	0,065	0,070	0,003	4,6
n-C14	0,069	0,068	0,067	0,061	0,066	0,003	5,2
n-C15	0,066	0,062	0,076	0,066	0,067	0,006	8,6
n-Ci6	0,049	0,051	0,050	0,048	0,049	0,001	2,0
n-C17	0,043	0,044	0,042	0,040	0,042	0,002	3,7
P	0,016	0,016	0,015	0,013	0,015	0,001	8,9
n-C18	0,038	0,038	0,037	0,033	0,037	0,003	6,9
F	0,007	0,008	0,007	0,007	0,007	0,000	6,3
n-C19	0,039	0,040	0,039	0,036	0,038	0,001	3,8
n-C20	0,041	0,041	0,041	0,039	0,040	0,001	2,6
n-C21	0,046	0,047	0,045	0,047	0,046	0,001	2,3
n-C22	0,042	0,044	0,042	0,043	0,043	0,001	2,4
n-C23	0,048	0,049	0,048	0,050	0,049	0,001	1,9
n-C24	0,038	0,041	0,040	0,042	0,040	0,002	4,1
n-C25	0,040	0,040	0,040	0,042	0,040	0,001	3,1
n-C26	0,031	0,035	0,033	0,039	0,034	0,003	10,0
n-C27	0,030	0,032	0,027	0,036	0,031	0,004	12,6
n-C28	0,021	0,024	0,024	0,028	0,024	0,003	11,3
n-C29	0,019	0,022	0,020	0,023	0,021	0,002	9,4
n-C30	0,017	0,018	0,016	0,019	0,017	0,001	7,2
T19	0,013	0,016	0,016	0,016	0,015	0,002	10,5
T20	0,019	0,021	0,025	0,022	0,022	0,002	11,1
T2I	0,015	0,018	0,015	0,014	0,016	0,002	10,9
T23	0,016	0,017	0,020	0,016	0,017	0,002	12,7

	AC1	AC2	AC3	AC4	Prom	SDs	RSDs
T24	0,011	0,013	0,014	0,011	0,012	0,001	10,2
T25	0,003	0,003	0,003	0,004	0,003	0,000	6,1
T26(R)	0,037	0,038	0,037	0,034	0,037	0,002	4,5
T26(S)	0,002	0,003	0,003	0,002	0,002	0,000	5,5
Ts	0,030	0,034	0,038	0,029	0,033	0,004	12,3
Tm	0,071	0,075	0,082	0,061	0,072	0,009	11,9
H29	0,291	0,266	0,244	0,256	0,264	0,020	7,6
H30	0,293	0,310	0,308	0,320	0,308	0,011	3,7
M30	0,026	0,023	0,023	0,028	0,025	0,003	10,5
H31 (S)	0,054	0,052	0,055	0,063	0,056	0,005	8,6
H31(R)	0,029	0,030	0,030	0,033	0,030	0,002	6,0
G30	0,009	0,007	0,008	0,008	0,008	0,001	10,0
H32 (S)	0,036	0,034	0,033	0,039	0,036	0,003	7,7
H32(R)	0,019	0,016	0,020	0,019	0,018	0,001	7,7
H33 (S)	0,014	0,012	0,016	0,014	0,014	0,001	10,4
H33(R)	0,011	0,009	0,010	0,008	0,010	0,001	12,8
S20	0,026	0,032	0,032	0,029	0,030	0,003	10,4
S21	0,038	0,042	0,044	0,035	0,040	0,004	10,0
S22	0,029	0,033	0,030	0,026	0,030	0,003	9,9
D27(βs)	0,082	0,088	0,092	0,084	0,087	0,005	5,4
D27(βr)	0,045	0,040	0,042	0,039	0,041	0,002	5,7
D27(αs)	0,011	0,009	0,012	0,011	0,011	0,001	12,0
D27(ar)	0,012	0,009	0,011	0,010	0,011	0,001	12,3
S27(as)	0,058	0,052	0,064	0,054	0,057	0,005	9,2
S27(βr)	0,083	0,073	0,087	0,085	0,082	0,006	7,6
S27(βs)	0,045	0,051	0,055	0,052	0,051	0,004	8,0
S27(ar)	0,054	0,070	0,064	0,056	0,061	0,007	11,5
S28(as)	0,001	0,001	0,001	0,001	0,001	0,000	11,7
S28(βr)	0,021	0,026	0,028	0,024	0,025	0,003	11,4
S28(βs)	0,029	0,034	0,032	0,030	0,031	0,002	7,8
S28(ar)	0,070	0,070	0,064	0,067	0,068	0,003	4,1
S29(as)	0,157	0,140	0,142	0,153	0,148	0,008	5,4
S29(βr)	0,062	0,064	0,053	0,059	0,059	0,005	8,4
S29(βs)	0,017	0,017	0,017	0,020	0,018	0,001	8,1
S29 (ar)	0,161	0,148	0,130	0,164	0,151	0,016	10,3
N	0,356	0,290	0,307	0,312	0,316	0,028	8,9
2-MN	0,078	0,089	0,086	0,086	0,085	0,005	5,4
1-MN	0,058	0,056	0,054	0,051	0,055	0,003	5,5
2-EN	0,012	0,014	0,014	0,012	0,013	0,001	7,6
1-EN	0,008	0,010	0,009	0,009	0,009	0,001	11,1
2.6 + 2.7 - DMN	0,066	0,067	0,065	0,064	0,066	0,001	1,9
1.3 + 1.7 - DMN	0,077	0,072	0,070	0,068	0,072	0,004	5,8
1,6 - DMN	0,037	0,044	0,043	0,041	0,041	0,003	7,5
1,4 + 2,3 - DMN	0,020	0,025	0,024	0,022	0,023	0,002	10,2
1,5 - DMN	0,010	0,011	0,011	0,010	0,011	0,001	5,6
1,2 - DMN	0,018	0,022	0,021	0,021	0,020	0,001	7,1
1,3,7 - TMN	0,028	0,032	0,031	0,029	0,030	0,002	6,0
1,3,6 - TMN	0,033	0,042	0,041	0,040	0,039	0,004	10,2
1,3,5 + 1,4,6 - TMN	0,024	0,031	0,030	0,032	0,029	0,004	12,2
2,3,6 - TMN	0,010	0,011	0,011	0,011	0,011	0,001	4,6
1,2,7 + 1,6,7 - TMN	0,013	0,017	0,016	0,014	0,015	0,002	12,3
1,2,6 - TMN	0,003	0,003	0,003	0,004	0,003	0,000	13,3
1,2,4 - TMN	0,006	0,007	0,007	0,007	0,007	0,001	12,9
1,2,5 - TMN	0,034	0,041	0,040	0,042	0,039	0,004	9,2
Ph	0,028	0,034	0,033	0,033	0,032	0,003	9,0
3-MP	0,015	0,016	0,016	0,018	0,016	0,001	7,6
2-MP	0,017	0,018	0,018	0,022	0,019	0,002	12,1
9-MP	0,027	0,024	0,024	0,026	0,025	0,002	7,0
1-MP	0,013	0,014	0,014	0,016	0,015	0,001	9,0
D	0,002	0,002	0,002	0,002	0,002	0,000	9,7
4-MDBT	0,004	0,004	0,004	0,004	0,004	0,000	3,1
2 + 3 - MDBT	0,002	0,002	0,002	0,002	0,002	0,000	5,8
1 - MDBT	0,001	0,001	0,001	0,001	0,001	0,000	9,1

Idem Table 11.

Table AC. Diagnostic ratios obtained from crude oil AC.

RDs	AC1	AC2	AC3	AC4	Prom	SDs	RSDs
P/F	2,201	2,047	2,228	1,815	2,073	0,189	9,1
P/n-C17	0,379	0,356	0,350	0,329	0,353	0,021	5,9
F/n-C18	0,192	0,198	0,179	0,220	0,197	0,017	8,6
n-C29%-C?7	0,442	0,500	0,486	0,587	0,503	0,061	12,1
H29/H30	0,994	0,858	0,793	0,798	0,861	0,094	10,9

					Averages	SDs	RSDs
10 x G30/G30 **x** C30	0,087	0,074	0,074	0,089	0,081	0,008	10,0
M30/H30	0,290	0,219	0,247	0,257	0,253	0,029	11,5
% S27	31,70	32,98	36,61	32,31	33,40	2,204	6,6
% S28	15,92	17,71	17,00	16,05	16,67	0,843	5,1
% S29	52,38	49,31	46,39	51,64	49,93	2,697	5,4
D27/S27	0,623	0,590	0,583	0,583	0,595	0,019	3,2
IMP	0,690	0,723	0,733	0,784	0,732	0,039	5,3
Rc	0,820	0,837	0,843	0,871	0,843	0,021	2,5
% 4-MeDBT	55,05	58,14	57,76	55,83	56,70	1,492	2,6
% 2+3-MeDBT	30,73	28,68	28,80	29,43	29,41	0,937	3,2
% 1-MeDBT	14,22	13,18	13,44	14,75	13,89	0,721	5,2
DBT/Ph	0,061	0,062	0,062	0,061	0,062	0,001	0,9

Idem Table 12.

Table BI. Relative standard deviations obtained from the relative abundances of BI crude oil.

RAs	BI1	BI2	BI3		BI4	Averages	SDs	RSDs
n-C9	0,056	0,053	0,057	-	0,055	0,002	3,2	
n-C10	0,092	0,086	0,076	-	0,084	0,008	9,6	
n-C11	0,102	0,096	0,084	-	0,094	0,009		10,0
n-C12	0,094	0,086	0,097	-	0,092	0,006	6,3	
n-C13	0,085	0,083	0,094	-	0,087	0,006	7,0	
n-C14	0,077	0,077	0,083	-	0,079	0,004	4,6	
n-C15	0,068	0,070	0,074	-	0,071	0,003	3,9	
n-C16	0,062	0,063	0,067	-	0,064	0,002	3,5	
n-C17	0,054	0,055	0,056	-	0,055	0,001		1,7
P	0,011	0,011	0,011	-	0,011	0,000	2,9	
n-C18	0,045	0,049	0,050	-	0,048	0,002		5,1
F	0,005	0,005	0,006	-	0,005	0,000		1,6
n-C19	0,041	0,043	0,043	-	0,042	0,001	3,0	
n-C20	0,036	0,039	0,037	-	0,037	0,002		4,1
n-C21	0,031	0,033	0,033	-	0,032	0,001	3,6	
n-C22	0,029	0,031	0,028	-	0,029	0,001	4,5	
n-C23	0,023	0,025	0,023	-	0,024	0,001		4,1
n-C24	0,021	0,023	0,020	-	0,021	0,001	6,3	
n-C25	0,017	0,019	0,015	-	0,017	0,002		10,3
n-C26	0,014	0,016	0,013	-	0,014	0,001	9,9	
n-C27	0,011	0,012	0,009	-	0,011	0,001		13,0
n-C28	0,010	0,012	0,010	-	0,011	0,001	8,9	
n-C29	0,009	0,009	0,008	-	0,008	0,001	7,9	
n-C30	0,005	0,005	0,005	-	0,005	0,000	4,8	
T19	0,000	0,000	0,000	-	0,000	0,000		-
T20	0,000	0,000	0,000	-	0,000	0,000		-
T21	0,000	0,000	0,000	-	0,000	0,000		-
T23	0,000	0,000	0,000	-	0,000	0,000		-
T24	0,000	0,000	0,000	-	0,000	0,000		-
T25	0,000	0,000	0,000	-	0,000	0,000		-
T26(**R**)	0,000	0,000	0,000	-	0,000	0,000		-
T26(**S**)	0,000	0,000	0,000	-	0,000	0,000		-
Ts	0,000	0,000	0,000	-	0,000	0,000		-
Tm	0,225	0,210	0,193	-	0,210	0,016	7,5	
H29	0,403	0,379	0,391	-	0,391	0,012		3,1
H30	0,241	0,268	0,257	-	0,255	0,014	5,4	
M30	0,054	0,056	0,061	-	0,057	0,004	6,9	
H31(**S**)	0,033	0,039	0,043	-	0,038	0,005		12,4
H31 (**R**)	0,045	0,047	0,055	-	0,049	0,005		10,9
G30	0,000	0,000	0,000	-	0,000	0,000		-
H32(**S**)	0,000	0,000	0,000	-	0,000	0,000		-
H32 (**R**)	0,000	0,000	0,000	-	0,000	0,000		-
H33(**S**)	0,000	0,000	0,000	-	0,000	0,000		-
H33 (**R**)	0,000	0,000	0,000	-	0,000	0,000		-
S20	0,000	0,000	0,000	-	0,000	0,000		-
S21	0,000	0,000	0,000	-	0,000	0,000		-
S22	0,000	0,000	0,000	-	0,000	0,000		-
D27(**βs**)	0,000	0,000	0,000	-	0,000	0,000		-
D27(**βr**)	0,000	0,000	0,000	-	0,000	0,000		-
D27(**αs**)	0,000	0,000	0,000	-	0,000	0,000		-
D27(**ar**)	0,000	0,000	0,000	-	0,000	0,000		-
S27 (**as**)	0,000	0,000	0,000	-	0,000	0,000		-
S27(**βr**)	0,000	0,000	0,000	-	0,000	0,000		-
S27 (**βs**)	0,000	0,000	0,000	-	0,000	0,000		-
S27 (**ar**)	0,340	0,325	0,319	-	0,328	0,011	3,2	
S28 (**as**)	0,000	0,000	0,000	-	0,000	0,000		-
S28(**βr**)	0,000	0,000	0,000	-	0,000	0,000		-
S28 (**βs**)	0,000	0,000	0,000	-	0,000	0,000		-

s28 (ar)	0,377	0,362	0,377	-	0,372	0,009	2,3
s29 (as)	0,121	0,136	0,130	-	0,129	0,008	6,1
s29(βr)	0,000	0,000	0,000	-	0,000	0,000	-
s29 (βs)	0,000	0,000	0,000	-	0,000	0,000	-
s29 (αr)	0,162	0,177	0,174	-	0,171	0,008	4,4
N	0,144	0,146	0,124	-	0,138	0,012	8,8
2-MN	0,159	0,151	0,164	-	0,158	0,007	4,3
1-MN	0,111	0,109	0,107	-	0,109	0,002	1,7
2-EN	0,016	0,016	0,017	-	0,016	0,001	5,1
1-EN	0,010	0,012	0,011	-	0,011	0,001	8,6
2.6 + 2.7 - DMN	0,064	0,069	0,073	-	0,069	0,005	6,8
1.3 + 1.7 - DMN	0,058	0,053	0,058	-	0,057	0,003	5,0
1,6 - DMN	0,050	0,046	0,053	-	0,049	0,004	7,4
1,4 + 2,3 - DMN	0,038	0,038	0,044	-	0,040	0,004	9,1
1,5 - DMN	0,012	0,013	0,015	-	0,014	0,001	9,0
1,2 - DMN	0,025	0,028	0,028	-	0,027	0,002	6,7
1,3,7 - TMN	0,020	0,023	0,024	-	0,023	0,002	9,9
1,3,6 - TMN	0,040	0,040	0,039	-	0,040	0,000	1,1
1,3,5 + 1,4,6 - TMN	0,017	0,021	0,020	-	0,019	0,002	10,6
2,3,6 - TMN	0,011	0,010	0,013	-	0,011	0,001	10,6
1,2,7 + 1,6,7 - TMN	0,015	0,014	0,015	-	0,015	0,000	2,7
1,2,6 - TMN	0,006	0,006	0,007	-	0,006	0,000	7,9
1,2,4 - TMN	0,004	0,004	0,003	-	0,004	0,001	14,2
1,2,5 - TMN	0,038	0,035	0,037	-	0,037	0,002	5,2
Ph	0,063	0,061	0,050	-	0,058	0,007	12,1
3-MP	0,024	0,029	0,029	-	0,028	0,003	10,1
2-MP	0,034	0,035	0,029	-	0,033	0,003	9,4
9-MP	0,019	0,017	0,020	-	0,019	0,001	7,4
1-MP	0,013	0,012	0,011	-	0,012	0,001	9,0
D	0,004	0,004	0,003	-	0,004	0,000	10,2
4-MDBT	0,003	0,003	0,003	-	0,003	0,000	3,5
2 + 3 - MDBT	0,002	0,002	0,002	-	0,002	0,000	4,6
1 - MDBT	0,001	0,001	0,001	-	0,001	0,000	9,9

Idem Table 11.

Table BI. Diagnostic ratios obtained from raw BI.

RDs	BI1	BI2	BI3	BI4	Prom	SDs	RSDs
P/F	1,955	2,056	2,043	-	2,018	0,055	2,7
PZn-C_{17}	0,195	0,198	0,200	-	0,198	0,002	1,3
FZn-C_{18}	0,120	0,109	0,110	-	0,113	0,006	5,3
n-C_2 9/n-C_{17}	0,160	0,164	0,138	-	0,154	0,014	9,1
H29/H30	1,674	1,414	1,523	-	1,537	0,131	8,5
10 x G30/G30 x C30	0,223	0,210	0,239	-	0,224	0,015	6,5
M30/H30	0,000	0,000	0,000	-	0,000	0,000	#DIV/0!
% S_{27}	33,96	32,46	31,92	-	32,78	1,056	3,2
% S_{28}	37,74	36,25	37,74	-	37,24	0,861	2,3
% S_{29}	28,30	31,29	30,35	-	29,98	1,529	5,1
D27/S27	0,000	0,000	0,000	-	0,000	0,000	#DIV/0!
IMP	0,913	1,069	1,080	-	1,021	0,094	9,2
Rc	0,942	1,028	1,034	-	1,001	0,052	5,1
% 4-MeDBT	50,24	50,12	50,36	-	50,24	0,119	0,2
% 2+3-MeDBT	33,73	33,88	31,31	-	32,97	1,440	4,4
% 1-MeDBT	16,04	16,00	18,33	-	16,79	1,335	8,0
DBTZPh	0,060	0,063	0,063	-	0,062	0,002	3,0

Idem Table 12.

Table BM. Relative standard deviations obtained from the relative abundances of the BM crude oil.

RAs	BM1	BM2	BM3	BM4	Averages	SDs	RSDs
n-C9	0,121	0,112	-	-	0,117	0,007	5,9
n-C10	0,118	0,104	-	-	0,111	0,010	8,7
n-C11	0,108	0,102	-	-	0,105	0,004	3,8
n-C12	0,091	0,089	-	-	0,090	0,001	1,5
n-C13	0,078	0,079	-	-	0,078	0,001	0,9
n-C14	0,068	0,069	-	-	0,068	0,000	0,3
n-C15	0,058	0,061	-	-	0,060	0,002	3,9
n-Ci6	0,053	0,055	-	-	0,054	0,002	3,0
n-C17	0,045	0,048	-	-	0,047	0,002	3,8
P	0,009	0,009	-	-	0,009	0,000	2,3
n-C18	0,039	0,042	-	-	0,040	0,002	5,5
F	0,005	0,005	-	-	0,005	0,000	0,1
n-C19	0,034	0,037	-	-	0,035	0,002	6,4

n-C20	0,031	0,034	-	-	0,032	0,002	6,6
n-C21	0,028	0,029	-	-	0,028	0,001	4,8
n-C22	0,024	0,027	-	-	0,025	0,002	6,4
n-C23	0,020	0,022	-	-	0,021	0,001	7,0
n-C24	0,017	0,020	-	-	0,018	0,002	10,9
n-C25	0,015	0,016	-	-	0,016	0,000	1,8
n-C26	0,012	0,013	-	-	0,012	0,001	6,8
n-C27	0,008	0,008	-	-	0,008	0,000	0,3
n-C28	0,007	0,008	-	-	0,008	0,001	9,9
n-C29	0,006	0,006	-	-	0,006	0,000	0,7
n-C30	0,005	0,005	-	-	0,005	0,000	0,6
T19	0,000	0,000	-	-	0,000	0,000	-
T20	0,000	0,000	-	-	0,000	0,000	-
T2i	0,000	0,000	-	-	0,000	0,000	-
T23	0,000	0,000	-	-	0,000	0,000	-
T24	0,000	0,000	-	-	0,000	0,000	-
T25	0,000	0,000	-	-	0,000	0,000	-
T26(R)	0,000	0,000	-	-	0,000	0,000	-
T26(S)	0,000	0,000	-	-	0,000	0,000	-
Ts	0,000	0,000	-	-	0,000	0,000	-
Tm	0,244	0,237	-	-	0,240	0,005	2,0
H29	0,326	0,320	-	-	0,323	0,004	1,4
H30	0,291	0,292	-	-	0,291	0,000	0,1
M30	0,048	0,052	-	-	0,050	0,003	5,1
H31(S)	0,044	0,050	-	-	0,047	0,005	9,9
H31(R)	0,047	0,049	-	-	0,048	0,002	3,7
G30	0,000	0,000	-	-	0,000	0,000	-
H32(S)	0,000	0,000	-	-	0,000	0,000	-
H32(R)	0,000	0,000	-	-	0,000	0,000	-
H33(S)	0,000	0,000	-	-	0,000	0,000	-
H33(R)	0,000	0,000	-	-	0,000	0,000	-
S20	0,000	0,000	-	-	0,000	0,000	-
S21	0,000	0,000	-	-	0,000	0,000	-
S22	0,000	0,000	-	-	0,000	0,000	-
D27(βs)	0,000	0,000	-	-	0,000	0,000	-
D27(βr)	0,000	0,000	-	-	0,000	0,000	-
D27(αs)	0,000	0,000	-	-	0,000	0,000	-
D27(ar)	0,000	0,000	-	-	0,000	0,000	-
S27(as)	0,000	0,000	-	-	0,000	0,000	-
S27(βr)	0,000	0,000	-	-	0,000	0,000	-
S27(βs)	0,000	0,000	-	-	0,000	0,000	-
S27(ar)	0,314	0,338	-	-	0,326	0,017	-
S28(as)	0,000	0,000	-	-	0,000	0,000	-
S28(βr)	0,000	0,000	-	-	0,000	0,000	-
S28(βs)	0,000	0,000	-	-	0,000	0,000	-
S28(ar)	0,368	0,369	-	-	0,368	0,001	0,2
S29(as)	0,133	0,121	-	-	0,127	0,009	7,0
S29(βr)	0,000	0,000	-	-	0,000	0,000	-
S29(βs)	0,000	0,000	--	0,000		0,000	-
S29(αr)	0,185	0,173	-	-	0,179	0,009	4,9
N	0,154	0,149	-	-	0,151	0,004	2,4
2-MN	0,163	0,163	-	-	0,163	0,000	0,1
1-MN	0,123	0,122	-	-	0,122	0,001	0,7
2-EN	0,016	0,017	-	-	0,016	0,000	3,0
1-EN	0,011	0,011	-	-	0,011	0,000	1,5
2.6 + 2.7 - DMN	0,081	0,083	-	-	0,082	0,002	1,9
1.3 + 1.7 - DMN	0,070	0,071	-	-	0,070	0,001	1,0
1,6 - DMN	0,055	0,056	-	-	0,055	0,001	1,7
1,4 + 2,3 - DMN	0,039	0,040	-	-	0,040	0,000	0,7
1,5 - DMN	0,013	0,013	-	-	0,013	0,000	1,8
1,2 - DMN	0,028	0,028	-	-	0,028	0,000	0,5
1,3,7 - TMN	0,023	0,024	-	-	0,023	0,001	3,3
1,3,6 - TMN	0,045	0,047	-	-	0,046	0,001	2,1
1,3,5 + 1,4,6 - TMN	0,023	0,023	-	-	0,023	0,000	1,4
2,3,6 - TMN	0,011	0,011	-	-	0,011	0,000	0,7
1,2,7 + 1,6,7 - TMN	0,017	0,017	-	-	0,017	0,000	1,1
1,2,6 - TMN	0,008	0,008	-	-	0,008	0,000	5,3
1,2,4 - TMN	0,005	0,006	-	-	0,006	0,000	1,5
1,2,5 - TMN	0,045	0,045	-	-	0,045	0,000	0,0
Ph	0,027	0,027	-	-	0,027	0,000	0,3
3-MP	0,010	0,009	-	-	0,010	0,001	5,5
2-MP	0,012	0,012	-	-	0,012	0,000	1,0

					Prom	SDs	RSDs
9-MP	0,009	0,008	-	-	0,008	0,001	11,7
1-MP	0,005	0,005	-	-	0,005	0,000	5,2
D	0,003	0,003	-	-	0,003	0,000	10,8
4-MDBT	0,002	0,002	-	-	0,002	0,000	4,7
2 + 3 - MDBT	0,001	0,001	-	-	0,001	0,000	2,9
1 - MDBT	0,001	0,001	-	-	0,001	0,000	8,8

Idem Table 11.

Table BM. Diagnostic ratios obtained from BM crude oil.

RDs	BM1	BM2	BM3	BM4	Prom	SDs	RSDs
PZF	1,848	1,905	-	-	1,877	0,040	2,1
P/n-C17	0,200	0,196	-	-	0,198	0,003	1,6
FZn-Cxs	0,127	0,117	-	-	0,122	0,007	5,4
n-C29/n-C17	0,136	0,127	-	-	0,132	0,006	4,6
H29/H30	1,121	1,098	-	-	1,109	0,016	1,5
10 x G3oG3o x C30	0,165	0,177	-	-	0,171	0,009	5,0
M3o/H3o	0,000	0,000	-	-	0,000	0,000	-
% S27	31,36	33,77	-	-	32,56	1,702	5,2
% S28	36,80	36,90	-	-	36,85	0,065	0,2
% S29	31,84	29,34	-	-	30,59	1,766	5,8
D27^S 27	0,000	0,000	-	-	0,000	0,000	-
IMP	0,803	0,819	-	-	0,811	0,012	1,4
Rc	0,882	0,891	-	-	0,886	0,006	0,7
% 4-MeDBT	51,83	51,94	-	-	51,89	0,075	0,1
% 2+3-MeDBT	30,85	31,69	-	-	31,27	0,600	1,9
% 1-MeDBT	17,32	16,37	-	-	16,84	0,675	4,0
DBTZPh	0,119	0,103	-	-	0,111	0,012	10,5

Idem Table 12.

Table BS. Relative standard deviation obtained from the relative abundances of the BS crude oil.

RAs	BS1	BS2	BS3	BS4	Averages	SDs	RSDs
n-C9	0,148	0,136	0,166	0,132	0,145	0,015	10,4
n-C10	0,145	0,137	0,141	0,127	0,138	0,008	5,8
n-C11	0,129	0,123	0,110	0,117	0,120	0,008	6,8
n-C12	0,098	0,097	0,089	0,096	0,095	0,004	4,2
n-Ci3	0,078	0,077	0,076	0,081	0,078	0,002	2,9
n-C14	0,060	0,062	0,061	0,066	0,063	0,003	4,2
n-C15	0,051	0,054	0,052	0,059	0,054	0,003	6,1
n-Ci6	0,048	0,046	0,048	0,051	0,048	0,002	3,8
n-C17	0,036	0,040	0,040	0,044	0,040	0,004	8,9
P	0,006	0,007	0,007	0,008	0,007	0,000	6,8
n-C18	0,031	0,035	0,034	0,037	0,034	0,003	7,9
F	0,004	0,004	0,004	0,004	0,004	0,000	7,1
n-C19	0,028	0,031	0,029	0,031	0,030	0,001	4,9
n-C20	0,025	0,028	0,026	0,029	0,027	0,002	6,5
n-C21	0,022	0,024	0,023	0,025	0,024	0,001	5,8
n-C22	0,020	0,022	0,020	0,021	0,020	0,001	4,3
n-C23	0,016	0,018	0,015	0,018	0,017	0,002	9,9
n-C24	0,014	0,016	0,016	0,015	0,015	0,001	7,1
n-C25	0,012	0,012	0,013	0,012	0,012	0,001	5,3
n-C26	0,009	0,010	0,010	0,008	0,009	0,001	8,7
n-C27	0,007	0,007	0,006	0,005	0,006	0,001	9,2
n-C28	0,006	0,006	0,006	0,005	0,005	0,000	7,7
n-C29	0,004	0,004	0,004	0,004	0,004	0,000	5,4
n-C30	0,005	0,005	0,004	0,004	0,004	0,000	8,9
T19	0,000	0,000	0,000	0,000	0,000	0,000	-
T20	0,000	0,000	0,000	0,000	0,000	0,000	-
T21	0,000	0,000	0,000	0,000	0,000	0,000	-
T23	0,000	0,000	0,000	0,000	0,000	0,000	-
T24	0,000	0,000	0,000	0,000	0,000	0,000	-
T25	0,000	0,000	0,000	0,000	0,000	0,000	-
T26(R)	0,000	0,000	0,000	0,000	0,000	0,000	-
T26(S)	0,000	0,000	0,000	0,000	0,000	0,000	-
Ts	0,000	0,000	0,000	0,000	0,000	0,000	-
Tm	0,161	0,150	0,147	0,150	0,152	0,006	4,1
H29	0,425	0,449	0,408	0,413	0,424	0,018	4,3
H30	0,286	0,274	0,308	0,300	0,292	0,015	5,1
M30	0,056	0,053	0,065	0,061	0,059	0,006	9,4
H31 (S)	0,034	0,035	0,037	0,041	0,037	0,003	7,9
H31(R)	0,038	0,039	0,034	0,035	0,036	0,002	6,0
G30	0,000	0,000	0,000	0,000	0,000	0,000	-
H32 (S)	0,000	0,000	0,000	0,000	0,000	0,000	-
H32(R)	0,000	0,000	0,000	0,000	0,000	0,000	-

H33 (S)	0,000	0,000	0,000	0,000	0,000	0,000	-
H33 (R)	0,000	0,000	0,000	0,000	0,000	0,000	-
S20	0,000	0,000	0,000	0,000	0,000	0,000	-
S21	0,000	0,000	0,000	0,000	0,000	0,000	-
S22	0,000	0,000	0,000	0,000	0,000	0,000	-
D27(βs)	0,000	0,000	0,000	0,000	0,000	0,000	-
D27(βr)	0,000	0,000	0,000	0,000	0,000	0,000	-
D27 (as)	0,000	0,000	0,000	0,000	0,000	0,000	-
D27 (ar)	0,000	0,000	0,000	0,000	0,000	0,000	-
S27 (as)	0,000	0,000	0,000	0,000	0,000	0,000	-
S27(βr)	0,000	0,000	0,000	0,000	0,000	0,000	-
S27 (βs)	0,000	0,000	0,000	0,000	0,000	0,000	-
S27 (ar)	0,328	0,330	0,306	0,309	0,318	0,013	4,0
S28 (as)	0,000	0,000	0,000	0,000	0,000	0,000	-
S28(βr)	0,000	0,000	0,000	0,000	0,000	0,000	-
S28(βs)	0,000	0,000	0,000	0,000	0,000	0,000	-
S28 (ar)	0,376	0,356	0,369	0,384	0,371	0,012	3,2
S29 (as)	0,120	0,132	0,134	0,123	0,127	0,007	5,4
S29(βr)	0,000	0,000	0,000	0,000	0,000	0,000	-
S29 (βs)	0,000	0,000	0,000	0,000	0,000	0,000	-
S29 (αr)	0,176	0,182	0,192	0,185	0,184	0,007	3,5
N	0,166	0,181	0,163	0,167	0,169	0,008	4,7
2-MN	0,174	0,173	0,155	0,157	0,165	0,010	6,2
1-MN	0,132	0,133	0,114	0,125	0,126	0,009	7,2
2-EN	0,016	0,015	0,015	0,013	0,015	0,001	7,2
1-EN	0,012	0,011	0,013	0,011	0,012	0,001	6,1
2.6 + 2.7 - DMN	0,075	0,074	0,085	0,074	0,077	0,006	7,3
1.3 + 1.7 - DMN	0,065	0,064	0,070	0,067	0,066	0,003	3,8
1,6 - DMN	0,053	0,053	0,062	0,057	0,056	0,004	7,3
1,4 + 2,3 - DMN	0,036	0,035	0,037	0,038	0,037	0,001	2,8
1,5 - DMN	0,012	0,012	0,013	0,014	0,013	0,001	6,9
1,2 - DMN	0,027	0,026	0,028	0,030	0,028	0,002	6,4
1,3,7 - TMN	0,017	0,017	0,018	0,019	0,018	0,001	4,2
1,3,6 - TMN	0,042	0,038	0,048	0,046	0,044	0,004	10,3
1,3,5 + 1,4,6 - TMN	0,020	0,019	0,022	0,023	0,021	0,002	8,4
2,3,6 - TMN	0,010	0,011	0,011	0,011	0,011	0,000	4,6
1,2,7 + 1,6,7 - TMN	0,017	0,018	0,018	0,020	0,018	0,001	6,0
1,2,6 - TMN	0,009	0,009	0,010	0,011	0,010	0,001	9,1
1,2,4 - TMN	0,005	0,005	0,004	0,006	0,005	0,001	10,2
1,2,5 - TMN	0,044	0,041	0,041	0,048	0,044	0,003	8,0
Ph	0,026	0,023	0,029	0,025	0,026	0,002	9,5
3-MP	0,011	0,009	0,009	0,011	0,010	0,001	9,6
2-MP	0,012	0,013	0,012	0,010	0,012	0,001	8,7
9-MP	0,008	0,009	0,010	0,009	0,009	0,001	7,7
1-MP	0,005	0,006	0,006	0,006	0,006	0,000	7,7
D	0,003	0,003	0,003	0,003	0,003	0,000	8,1
4-MDBT	0,001	0,001	0,001	0,001	0,001	0,000	4,5
2 + 3 - MDBT	0,001	0,001	0,001	0,001	0,001	0,000	7,0
1 - MDBT	0,001	0,001	0,000	0,001	0,001	0,000	6,3

Idem Table 11.

Table BS. Diagnostic ratios obtained from BS crude oil.

RDs	BS1	BS2	BS3	BS4	Prom	SDs	RSDs
P/F	1,761	1,746	2,065	2,044	1,904	0,174	9,1
P/n-C17	0,181	0,180	0,185	0,171	0,179	0,006	3,3
FZn-Cjs	0,119	0,119	0,103	0,099	0,110	0,010	9,4
n-C29½-C17	0,111	0,101	0,112	0,098	0,106	0,007	7,0
H29/H30	1,485	1,639	1,328	1,378	1,458	0,138	9,5
10 x G30/G30 x C30	0,197	0,192	0,213	0,203	0,201	0,009	4,4
M30/H30	0,000	0,000	0,000	0,000	0,000	0,000	-
% S27	32,79	33,02	30,58	30,88	31,82	1,263	4,0
% S28	37,64	35,59	36,90	38,37	37,12	1,188	3,2
% S29	29,57	31,40	32,52	30,75	31,06	1,235	4,0
D27/S27	0,000	0,000	0,000	0,000	0,000	0,000	-
IMP	0,873	0,844	0,713	0,784	0,804	0,071	8,8
Rc	0,920	0,904	0,832	0,871	0,882	0,039	4,4
% 4-MeDBT	47,51	46,75	46,17	43,89	46,08	1,559	3,4
% 2+3-MeDBT	27,15	28,80	31,89	31,63	29,86	2,289	7,7
% 1-MeDBT	25,34	24,46	21,95	24,48	24,06	1,465	6,1
DBT/Ph	0,114	0,119	0,095	0,130	0,115	0,015	12,8

Idem Table 12.

Table CI. Relative standard deviations obtained from the relative abundances of CI crude oil.

RAs	CI1	CI2	CI3	CI4	Averages	SDs	RSDs
n-C9	0,081	0,075	-	-	0,078	0,005	5,9
n-C10	0,084	0,080	-	-	0,082	0,003	3,4
n-C11	0,081	0,078	-	-	0,079	0,002	2,3
n-C12	0,082	0,076	-	-	0,079	0,004	5,2
n-C13	0,074	0,071	-	-	0,072	0,002	2,7
n-C14	0,066	0,066	-	-	0,066	0,000	0,2
II-C15	0,063	0,063	-	-	0,063	0,000	0,2
n-C16	0,058	0,059	-	-	0,058	0,001	1,4
n-C17	0,053	0,054	-	-	0,054	0,000	0,9
P	0,024	0,023	-	-	0,024	0,000	1,7
n-C18	0,046	0,047	-	-	0,047	0,001	1,6
F	0,013	0,014	-	-	0,013	0,001	4,6
n-C19	0,041	0,045	-	-	0,043	0,002	5,8
n-C20	0,038	0,041	-	-	0,040	0,002	4,6
n-C21	0,037	0,037	-	-	0,037	0,000	1,1
n-C22	0,031	0,034	-	-	0,033	0,002	6,0
n-C23	0,026	0,029	-	-	0,027	0,002	8,7
n-C24	0,025	0,027	-	-	0,026	0,001	4,3
n-C25	0,021	0,022	-	-	0,022	0,001	3,8
n-C26	0,015	0,018	-	-	0,017	0,002	10,3
n-C27	0,012	0,012	-	-	0,012	0,000	1,8
n-C28	0,011	0,012	-	-	0,012	0,001	6,3
n-C29	0,009	0,010	-	-	0,009	0,000	2,8
n-C30	0,008	0,007	-	-	0,008	0,001	8,9
T19	0,000	0,000	-	-	0,000	0,000	-
T20	0,000	0,000	-	-	0,000	0,000	-
T21	0,000	0,000	-	-	0,000	0,000	-
T23	0,000	0,000	-	-	0,000	0,000	-
T24	0,000	0,000	-	-	0,000	0,000	-
T25	0,000	0,000	-	-	0,000	0,000	-
T26(R)	0,000	0,000	-	-	0,000	0,000	-
T26(S)	0,000	0,000	-	-	0,000	0,000	-
Ts	0,000	0,000	-	-	0,000	0,000	-
Tm	0,185	0,188	-	-	0,186	0,002	0,9
H29	0,323	0,308	-	-	0,316	0,011	3,4
H30	0,340	0,339	-	-	0,339	0,001	0,3
M30	0,041	0,045	-	-	0,043	0,003	6,2
H31(S)	0,070	0,074	-	-	0,072	0,003	3,6
H31(R)	0,040	0,047	-	-	0,044	0,005	10,8
G30	0,000	0,000	-	-	0,000	0,000	-
H32(S)	0,000	0,000	-	-	0,000	0,000	-
H32(R)	0,000	0,000	-	-	0,000	0,000	-
H33(S)	0,000	0,000	-	-	0,000	0,000	-
H33(R)	0,000	0,000	-	-	0,000	0,000	-
S20	0,281	0,283	-	-	0,282	0,002	0,7
S21	0,111	0,109	-	-	0,110	0,001	1,3
S22	0,057	0,057	-	-	0,057	0,000	0,5
D27(βs)	0,117	0,115	-	-	0,116	0,001	1,1
D27(βr)	0,063	0,060	-	-	0,062	0,002	3,9
D27(αs)	0,028	0,025	-	-	0,027	0,002	8,0
D27(ar)	0,017	0,018	-	-	0,017	0,001	3,3
S27(as)	0,033	0,029	-	-	0,031	0,003	10,6
S27(βr)	0,053	0,058	-	-	0,056	0,004	6,7
S27(βs)	0,044	0,049	-	-	0,047	0,003	6,8
S27(ar)	0,033	0,033	-	-	0,033	0,000	0,2
S28(as)	0,024	0,024	-	-	0,024	0,000	0,6
S28(βr)	0,018	0,020	-	-	0,019	0,001	7,6
S28(βs)	0,012	0,012	-	-	0,012	0,000	3,2
S28(ar)	0,043	0,045	-	-	0,044	0,001	2,7
S29(as)	0,022	0,020	-	-	0,021	0,001	6,7
S29(βr)	0,007	0,006	-	-	0,006	0,001	9,4
S29(βs)	0,010	0,010	--		0,010	0,000	1,5
S29(αr)	0,027	0,027	--		0,027	0,000	0,4
N	0,156	0,145	--		0,150	0,008	5,3
2-MN	0,115	0,109	--		0,112	0,004	3,8
1-MN	0,090	0,085	--		0,087	0,003	3,6
2-EN	0,024	0,028	--		0,026	0,002	8,4
1-EN	0,010	0,011	--		0,010	0,001	7,5
2.6 + 2.7 - DMN	0,102	0,098	--		0,100	0,003	2,7

1.3 + 1.7 - DMN	0,057	0,053	--	0,055	0,003	6,0
1,6 - DMN	0,039	0,038	--	0,038	0,001	3,1
1,4 + 2,3 - DMN	0,034	0,039	--	0,036	0,004	9,9
1,5 - DMN	0,008	0,009	--	0,008	0,001	9,5
1,2 - DMN	0,020	0,023	--	0,022	0,002	9,4
1,3,7 - TMN	0,022	0,023	--	0,022	0,001	6,1
1,3,6 - TMN	0,032	0,035	--	0,034	0,002	6,7
1,3,5 + 1,4,6 - TMN	0,023	0,027	--	0,025	0,002	9,3
2,3,6 - TMN	0,013	0,013	--	0,013	0,000	1,9
1,2,7 + 1,6,7 - TMN	0,012	0,012	--	0,012	0,000	2,7
1,2,6 - TMN	0,009	0,009	--	0,009	0,000	1,2
1,2,4 - TMN	0,003	0,003	--	0,003	0,000	7,8
1,2,5 - TMN	0,018	0,021	--	0,019	0,002	8,9
Ph	0,072	0,070	--	0,071	0,001	1,3
3-MP	0,044	0,046	--	0,045	0,002	3,8
2-MP	0,049	0,052	--	0,051	0,002	3,2
9-MP	0,019	0,021	--	0,020	0,001	6,4
1-MP	0,014	0,015	--	0,014	0,001	5,8
D	0,007	0,007	--	0,007	0,000	3,4
4-MDBT	0,004	0,005	--	0,005	0,000	6,9
2 + 3 - MDBT	0,002	0,003	--	0,002	0,000	8,7
1 - MDBT	0,001	0,001	--	0,001	0,000	7,3

Idem Table 11.

Table CI. Diagnostic ratios obtained from raw CI.

RDs	CI1	CI2	CI3	CI4	Prom	SDs	RSDs
P/F	1,877	1,718	-	-	1,797	0,112	6,2
P/n-C17	0,448	0,432	-	-	0,440	0,011	2,6
F⅛-C 8₁	0,275	0,287	-	-	0,281	0,008	2,9
n-C29⅛-C17	0,173	0,177	-	-	0,175	0,003	1,9
H29/H30	0,952	0,910	-	-	0,931	0,029	3,2
10 x G30/G30 x C30	0,122	0,134	-	-	0,128	0,008	6,4
M30/H30	0,000	0,000	-	-	0,000	0,000	-
% S27	50,17	50,74	-	-	50,46	0,405	0,8
% S28	29,58	30,37	-	-	29,97	0,555	1,9
% S29	20,25	18,89	-	-	19,57	0,960	4,9
D27/S27	1,373	1,288	-	-	1,330	0,060	4,5
IMP	1,332	1,377	-	-	1,355	0,032	2,4
Rc	1,173	1,197	-	-	1,185	0,018	1,5
% 4-MeDBT	58,15	57,64	-	-	57,89	0,356	0,6
% 2+3-MeDBT	30,48	31,01	-	-	30,75	0,373	1,2
% 1-MeDBT	11,37	11,35	-	-	11,36	0,018	0,2
DBT/Ph	0,091	0,098	-	-	0,095	0,004	4,7

Idem Table 12.

Table CO. Relative standard deviation obtained from the relative abundances of crude CO.

RAs	CO1	CO2	CO3	CO4	Averages	SDs	RSDs
n-C9	0,079	0,072	0,066	-	0,072	0,007	9,6
n-C10	0,081	0,079	0,071	-	0,077	0,006	7,2
n-C11	0,086	0,083	0,077	-	0,082	0,005	5,6
n-C12	0,077	0,078	0,084	-	0,080	0,003	4,3
n-C13	0,073	0,072	0,076	-	0,073	0,002	3,1
n-C14	0,064	0,065	0,066	-	0,065	0,001	1,3
n-C15	0,061	0,064	0,064	-	0,063	0,001	2,2
n-C16	0,061	0,058	0,059	-	0,059	0,001	2,5
n-C17	0,054	0,054	0,054	-	0,054	0,000	0,3
P	0,021	0,022	0,021	-	0,021	0,001	3,4
n-C18	0,045	0,047	0,048	-	0,047	0,002	3,9
F	0,012	0,013	0,014	-	0,013	0,001	9,1
n-C19	0,044	0,044	0,044	-	0,044	0,000	0,9
n-C20	0,040	0,041	0,040	-	0,040	0,001	1,3
n-C21	0,035	0,037	0,038	-	0,037	0,001	3,1
n-C22	0,032	0,034	0,035	-	0,034	0,002	5,3
n-C23	0,028	0,028	0,031	-	0,029	0,001	4,7
n-C24	0,027	0,027	0,027	-	0,027	0,000	1,3
n-C25	0,022	0,023	0,024	-	0,023	0,001	4,8
n-C26	0,016	0,018	0,018	-	0,017	0,001	6,9
n-C27	0,012	0,012	0,013	-	0,012	0,001	4,1
n-C28	0,012	0,013	0,013	-	0,013	0,000	3,6
n-C29	0,010	0,010	0,010	-	0,010	0,000	1,0
n-C30	0,007	0,007	0,008	-	0,007	0,000	6,4
T19	0,000	0,000	0,000	-	0,000	0,000	-

	CO1	CO2	CO3	CO4	Prom	SDs	RSDs
T_{20}	0,000	0,000	0,000	-	0,000	0,000	-
T_{21}	0,000	0,000	0,000	-	0,000	0,000	-
T_{23}	0,000	0,000	0,000	-	0,000	0,000	-
T_{24}	0,000	0,000	0,000	-	0,000	0,000	-
T_{25}	0,000	0,000	0,000	-	0,000	0,000	-
$T_{26}(R)$	0,000	0,000	0,000	-	0,000	0,000	-
$T_{26}(S)$	0,000	0,000	0,000	-	0,000	0,000	-
Ts	0,000	0,000	0,000	-	0,000	0,000	
Tm	0,161	0,182	0,153	-	0,165	0,015	9,1
H_{29}	0,323	0,289	0,293	-	0,302	0,019	6,2
H_{30}	0,363	0,373	0,364	-	0,367	0,005	1,5
M_{30}	0,054	0,051	0,059	-	0,055	0,004	7,7
$H_{31}(S)$	0,056	0,064	0,084	-	0,068	0,015	21,5
$H_{31}(R)$	0,044	0,041	0,047	-	0,044	0,003	6,9
G_{30}	0,000	0,000	0,000	-	0,000	0,000	-
$H_{32}(S)$	0,000	0,000	0,000	-	0,000	0,000	-
$H_{32}(R)$	0,000	0,000	0,000	-	0,000	0,000	-
$H_{33}(S)$	0,000	0,000	0,000	-	0,000	0,000	-
$H_{33}(R)$	0,000	0,000	0,000	-	0,000	0,000	-
S_{20}	0,321	0,306	0,302	-	0,310	0,010	3,2
S_{21}	0,099	0,110	0,098	-	0,102	0,007	6,4
S_{22}	0,044	0,047	0,053	-	0,048	0,005	9,7
$D_{27}(\beta s)$	0,098	0,105	0,097	-	0,100	0,004	4,2
$D_{27}(\beta r)$	0,072	0,067	0,063	-	0,067	0,004	6,7
$D_{27}(\alpha s)$	0,015	0,015	0,015	-	0,015	0,000	1,6
$D_{27}(ar)$	0,023	0,021	0,022	-	0,022	0,001	5,1
$S_{27}(as)$	0,038	0,036	0,032	-	0,036	0,003	9,3
$S_{27}(\beta r)$	0,042	0,045	0,050	-	0,046	0,004	7,9
$S_{27}(\beta s)$	0,039	0,039	0,044	-	0,041	0,003	7,1
$S_{27}(ar)$	0,027	0,028	0,029	-	0,028	0,001	3,4
$S_{28}(as)$	0,029	0,025	0,024	-	0,026	0,002	8,6
$S_{28}(\beta r)$	0,021	0,022	0,025	-	0,023	0,002	8,2
$S_{28}(\beta s)$	0,014	0,012	0,012	-	0,013	0,001	7,6
$S_{28}(ar)$	0,042	0,045	0,051	-	0,046	0,005	10,9
$S_{29}(as)$	0,025	0,026	0,029	-	0,027	0,002	8,2
$S_{29}(\beta r)$	0,007	0,008	0,008	-	0,007	0,001	9,0
$S_{29}(\beta s)$	0,014	0,013	0,016	-	0,014	0,001	10,2
$S_{29}(\alpha r)$	0,031	0,030	0,030	-	0,030	0,001	2,5
N	0,156	0,151	0,150	-	0,152	0,004	2,3
2-MN	0,117	0,113	0,123	-	0,118	0,005	3,9
1-MN	0,089	0,085	0,104	-	0,093	0,010	10,5
2-EN	0,023	0,022	0,025	-	0,023	0,002	6,9
1-EN	0,010	0,010	0,012	-	0,011	0,001	10,6
2.6 + 2.7 - DMN	0,094	0,089	0,080	-	0,088	0,007	8,1
1.3 + 1.7 - DMN	0,055	0,052	0,060	-	0,056	0,004	7,4
1,6 - DMN	0,038	0,037	0,042	-	0,039	0,003	7,1
1,4 + 2,3 - DMN	0,032	0,031	0,033	-	0,032	0,001	4,5
1,5 - DMN	0,006	0,006	0,005	-	0,006	0,001	9,9
1,2 - DMN	0,019	0,018	0,018	-	0,018	0,001	2,9
1,3,7 - TMN	0,016	0,015	0,014	-	0,015	0,001	5,6
1,3,6 - TMN	0,028	0,027	0,029	-	0,028	0,001	4,2
1,3,5 + 1,4,6 - TMN	0,010	0,009	0,009	-	0,009	0,000	3,3
2,3,6 - TMN	0,010	0,010	0,011	-	0,010	0,001	6,0
1,2,7 + 1,6,7 - TMN	0,011	0,011	0,012	-	0,011	0,000	3,5
1,2,6 - TMN	0,001	0,001	0,002	-	0,001	0,000	8,7
1,2,4 - TMN	0,003	0,003	0,002	-	0,003	0,000	10,7
1,2,5 - TMN	0,019	0,019	0,016	-	0,018	0,002	10,9
Ph	0,091	0,108	0,090	-	0,097	0,010	10,5
3-MPh	0,054	0,057	0,051	-	0,054	0,003	5,7
2-MPh	0,058	0,066	0,059	-	0,061	0,004	7,1
9-MPh	0,029	0,028	0,025	-	0,027	0,002	7,4
1-MPh	0,019	0,019	0,017	-	0,018	0,001	5,6
D	0,005	0,006	0,005	-	0,005	0,001	9,4
4-MDBT	0,003	0,004	0,003	-	0,004	0,000	8,4
2 + 3 - MDBT	0,002	0,002	0,002	-	0,002	0,000	9,3
1 - MDBT	0,001	0,001	0,001	-	0,001	0,000	8,6

Idem Table 11.

Table CO. Diagnostic ratios obtained from crude CO.

RDs	CO1	CO2	CO3	CO4	Prom	SDs	RSDs

P/F	1,778	1,704	1,524	-	1,669	0,131	7,8
P/n-C17	0,380	0,405	0,388	-	0,391	0,013	3,3
FZn-C$_{28}$	0,259	0,276	0,287	-	0,274	0,014	5,2
Π-C29/Π-C17	0,193	0,190	0,188	-	0,191	0,002	1,3
H$_2$ 9/H30	0,890	0,776	0,804	-	0,823	0,059	7,2
10 x G3o/G3o X C30	0,148	0,137	0,163	-	0,149	0,013	8,7
M3o/H3o	0,000	0,000	0,000	-	0,000	0,000	-
% S$_{27}$	44,50	45,11	44,08	-	44,56	0,518	1,2
% S28	32,06	31,73	32,23	-	32,01	0,255	0,8
% S29	23,44	23,16	23,69	-	23,43	0,264	1,1
D27/S27	1,420	1,396	1,281	-	1,366	0,074	5,4
IMP	1,207	1,194	1,245	-	1,215	0,026	2,2
Rc	1,104	1,097	1,125	-	1,108	0,015	1,3
% 4-MeDBT	55,00	54,38	54,17	-	54,52	0,431	0,8
% 2+3-MeDBT	31,82	32,20	31,92	-	31,98	0,196	0,6
% 1-MeDBT	13,18	13,42	13,91	-	13,50	0,369	2,7
DBTZPh	0,055	0,055	0,058	-	0,056	0,001	2,5

Idem Table 12.

Table DA. Relative standard deviation obtained from the relative abundances of crude DA.

RAs	DA1	DA2	DA3	DA4	Averages	SDs	RSDs
n-C9	0,044	0,046	0,040	0,049	0,045	0,004	8,9
n-C10	0,055	0,055	0,060	0,064	0,058	0,004	7,0
n-C11	0,058	0,056	0,063	0,052	0,057	0,005	8,4
n-C12	0,061	0,061	0,068	0,069	0,065	0,004	6,8
n-C13	0,053	0,056	0,057	0,054	0,055	0,002	3,9
n-C14	0,052	0,055	0,058	0,058	0,055	0,003	5,5
n-C15	0,061	0,056	0,061	0,060	0,060	0,002	3,9
n-C16	0,055	0,055	0,055	0,058	0,056	0,002	2,9
n-C17	0,051	0,056	0,055	0,057	0,055	0,002	4,3
P	0,039	0,038	0,039	0,038	0,039	0,000	1,1
n-C18	0,049	0,049	0,050	0,049	0,049	0,000	0,7
F	0,021	0,023	0,022	0,024	0,022	0,001	6,3
n-C19	0,051	0,054	0,051	0,051	0,052	0,001	2,8
n-C20	0,045	0,048	0,046	0,046	0,046	0,001	2,1
n-C21	0,051	0,046	0,045	0,046	0,047	0,003	5,7
n-C22	0,043	0,044	0,042	0,043	0,043	0,001	2,2
n-C23	0,043	0,041	0,037	0,036	0,039	0,003	8,0
n-C24	0,039	0,039	0,035	0,032	0,036	0,003	9,4
n-C25	0,036	0,036	0,030	0,030	0,033	0,004	10,6
n-C26	0,032	0,028	0,028	0,027	0,029	0,002	7,5
n-C27	0,023	0,020	0,019	0,020	0,021	0,002	8,8
n-C28	0,018	0,016	0,017	0,015	0,017	0,002	9,6
n-C29	0,013	0,012	0,015	0,012	0,013	0,001	9,9
n-C30	0,008	0,008	0,008	0,010	0,009	0,001	8,8
T19	0,094	0,090	0,097	0,085	0,092	0,005	5,7
T20	0,141	0,150	0,127	0,138	0,139	0,009	6,8
T21	0,268	0,280	0,259	0,280	0,272	0,010	3,6
T23	0,217	0,195	0,205	0,214	0,208	0,010	4,8
T24	0,156	0,159	0,175	0,159	0,162	0,009	5,3
T25	0,000	0,000	0,000	0,000	0,000	0,000	-
T26 (R)	0,077	0,076	0,084	0,068	0,076	0,007	8,6
T26(S)	0,045	0,050	0,053	0,057	0,051	0,005	9,1
Ts	0,000	0,000	0,000	0,000	0,000	0,000	-
Tm	0,000	0,000	0,000	0,000	0,000	0,000	-
H29	0,000	0,000	0,000	0,000	0,000	0,000	-
H30	0,000	0,000	0,000	0,000	0,000	0,000	-
M30	0,000	0,000	0,000	0,000	0,000	0,000	-
H31(S)	0,000	0,000	0,000	0,000	0,000	0,000	-
H31 (R)	0,000	0,000	0,000	0,000	0,000	0,000	-
G30	0,000	0,000	0,000	0,000	0,000	0,000	-
H32(S)	0,000	0,000	0,000	0,000	0,000	0,000	-
H32 (R)	0,000	0,000	0,000	0,000	0,000	0,000	-
H33(S)	0,000	0,000	0,000	0,000	0,000	0,000	-
H33 (R)	0,000	0,000	0,000	0,000	0,000	0,000	-
S20	0,254	0,250	0,213	0,233	0,237	0,019	7,9
S21	0,138	0,138	0,141	0,142	0,140	0,002	1,6
S22	0,109	0,105	0,113	0,104	0,108	0,004	4,0
D27(βs)	0,134	0,132	0,147	0,132	0,136	0,007	5,1
D27(βr)	0,054	0,055	0,057	0,052	0,054	0,002	4,0
D27 (as)	0,018	0,017	0,018	0,016	0,017	0,001	6,8
D27 (ar)	0,014	0,015	0,016	0,018	0,016	0,002	11,6
S27 (as)	0,018	0,020	0,018	0,019	0,019	0,001	3,3
S27(βr)	0,073	0,079	0,072	0,080	0,076	0,004	5,6
S27 (βs)	0,055	0,055	0,058	0,059	0,057	0,002	4,0
S27 (ar)	0,016	0,014	0,015	0,016	0,015	0,001	5,7

s28 (as)	0,002	0,002	0,002	0,002	0,002	0,000	9,2
s28(βr)	0,033	0,037	0,041	0,037	0,037	0,003	8,6
s28(βs)	0,008	0,009	0,010	0,008	0,009	0,001	11,4
s28 (ar)	0,024	0,022	0,025	0,028	0,025	0,002	9,5
s29 (as)	0,017	0,018	0,019	0,017	0,018	0,001	4,4
s29(βr)	0,010	0,010	0,011	0,013	0,011	0,001	11,7
s29 (βs)	0,008	0,007	0,007	0,006	0,007	0,001	12,0
s29 (αr)	0,017	0,017	0,017	0,019	0,017	0,001	5,0
N	0,090	0,086	0,076	0,095	0,087	0,008	9,5
2-MN	0,119	0,108	0,101	0,121	0,112	0,009	8,1
1-MN	0,065	0,057	0,059	0,071	0,063	0,006	10,2
2-EN	0,017	0,018	0,019	0,021	0,019	0,002	8,8
1-EN	0,000	0,000	0,000	0,000	0,000	0,000	8,1
2.6 + 2.7 - DMN	0,119	0,112	0,137	0,130	0,125	0,011	8,9
1.3 + 1.7 - DMN	0,036	0,042	0,047	0,043	0,042	0,005	11,0
1,6 - DMN	0,032	0,031	0,030	0,030	0,031	0,001	2,9
1,4 + 2,3 - DMN	0,042	0,042	0,038	0,039	0,040	0,002	4,5
1,5 - DMN	0,004	0,004	0,005	0,004	0,004	0,000	11,2
1,2 - DMN	0,003	0,003	0,003	0,003	0,003	0,000	2,5
1,3,7 - TMN	0,042	0,037	0,037	0,038	0,038	0,002	6,1
1,3,6 - TMN	0,048	0,046	0,046	0,043	0,046	0,002	4,7
1,3,5 + 1,4,6 - TMN	0,004	0,004	0,004	0,003	0,004	0,000	7,1
2,3,6 - TMN	0,037	0,042	0,036	0,036	0,038	0,003	8,0
1,2,7 + 1,6,7 - TMN	0,000	0,000	0,000	0,000	0,000	0,000	-
1,2,6 - TMN	0,000	0,000	0,000	0,000	0,000	0,000	-
1,2,4 - TMN	0,000	0,000	0,000	0,000	0,000	0,000	11,3
1,2,5 - TMN	0,001	0,001	0,001	0,001	0,001	0,000	11,2
Ph	0,084	0,083	0,085	0,070	0,080	0,007	8,5
3-MPh	0,067	0,072	0,081	0,060	0,070	0,009	12,6
2-MPh	0,064	0,069	0,073	0,056	0,065	0,008	11,6
9-MPh	0,043	0,049	0,042	0,045	0,045	0,003	7,2
1-MPh	0,037	0,043	0,034	0,040	0,039	0,004	10,5
D	0,011	0,013	0,011	0,011	0,011	0,001	6,9
4-MDBT	0,025	0,025	0,022	0,026	0,025	0,002	8,0
2 + 3 - MDBT	0,012	0,014	0,012	0,014	0,013	0,001	8,7
1 - MDBT	0,000	0,000	0,000	0,000	0,000	0,000	8,7

Idem Table 11.

Table DA. Diagnostic relationships obtained from crude DA.

RDs	DA1	DA2	DA3	DA4	Prom	SDs	RSDs
P/F	1,903	1,678	1,758	1,607	1,737	0,127	7,3
P⅛-C$_{17}$	0,760	0,677	0,704	0,677	0,704	0,039	5,6
F⅛-C 8$_1$	0,417	0,461	0,440	0,484	0,451	0,029	6,4
n-C29⅛-C17	0,253	0,212	0,268	0,212	0,236	0,028	12,0
H29/H30	-	-	-	-	-	-	-
10 x G30/G30 x C30	-	-	-	-	-	-	-
M30/H30	-	-	-	-	-	-	-
% s27	57,71	57,90	55,29	57,35	57,06	1,207	2,1
% s28	23,89	24,33	26,40	24,76	24,85	1,094	4,4
% s29	18,40	17,76	18,32	17,88	18,09	0,315	1,7
D27/S27	1,363	1,312	1,452	1,257	1,346	0,083	6,2
IMP	1,204	1,205	1,432	1,113	1,238	0,136	11,0
Rc	1,102	1,103	1,228	1,052	1,121	0,075	6,7
% 4-MeDBT	67,24	63,75	62,89	64,49	64,59	1,883	2,9
% 2+3-MeDBT	31,71	35,33	36,10	34,46	34,40	1,916	5,6
% 1-MeDBT	1,050	0,923	1,008	1,054	1,009	0,061	6,0
DBT/Ph	0,136	0,152	0,131	0,152	0,143	0,011	7,7

Idem Table 12.

Table DB. Relative standard deviation obtained from the relative abundances of crude DB.

RAs	DB1	DB2	DB3	DB4	Averages	SDs	RSDs
n-C9	0,035	0,039	0,036	-	0,036	0,002	4,9
n-C10	0,043	0,044	0,042	-	0,043	0,001	2,4
n-C11	0,052	0,052	0,045	-	0,050	0,004	8,1
n-C12	0,051	0,050	0,049	-	0,050	0,001	1,8
n-Ci3	0,048	0,049	0,048	-	0,048	0,000	1,0
n-C14	0,049	0,050	0,047	-	0,049	0,002	3,2
n-C15	0,060	0,055	0,055	-	0,057	0,003	5,4
n-Ci6	0,057	0,056	0,055	-	0,056	0,001	2,2
n-C17	0,055	0,058	0,058	-	0,057	0,002	3,1

P	0,038	0,039	0,041	-	0,039	0,002	4,7
n-C18	0,050	0,053	0,053	-	0,052	0,002	3,3
F	0,023	0,024	0,025	-	0,024	0,001	4,5
n-C19	0,056	0,058	0,058	-	0,057	0,001	1,9
П-C20	0,051	0,052	0,056	-	0,053	0,002	4,2
n-C21	0,051	0,050	0,053	-	0,051	0,001	2,6
n-C22	0,051	0,049	0,050	-	0,050	0,001	2,0
n-C23	0,046	0,043	0,046	-	0,045	0,001	3,1
n-C24	0,045	0,043	0,042	-	0,043	0,001	3,4
П-C25	0,041	0,040	0,041	-	0,041	0,000	1,1
П-C26	0,031	0,032	0,033	-	0,032	0,001	2,4
П-C27	0,022	0,022	0,026	-	0,023	0,002	9,7
П-C28	0,023	0,020	0,022	-	0,022	0,001	6,1
П-C29	0,015	0,014	0,013	-	0,014	0,001	5,5
n-C30	0,008	0,009	0,008	-	0,008	0,001	6,9
T19	0,169	0,157	0,164	-	0,163	0,006	3,9
T20	0,137	0,130	0,152	-	0,140	0,012	8,3
T2i	0,243	0,261	0,265	-	0,256	0,012	4,5
T23	0,212	0,203	0,179	-	0,198	0,017	8,6
T24	0,122	0,129	0,117	-	0,123	0,006	4,9
T25	0,000	0,000	0,000	-	0,000	0,000	-
T26(R)	0,066	0,072	0,075	-	0,071	0,004	6,3
T26(S)	0,051	0,049	0,048	-	0,049	0,001	3,0
Ts	0,000	0,000	0,000	-	0,000	0,000	-
Tm	0,000	0,000	0,000	-	0,000	0,000	-
H29	0,000	0,000	0,000	-	0,000	0,000	-
H30	0,000	0,000	0,000	-	0,000	0,000	-
M30	0,000	0,000	0,000	-	0,000	0,000	-
H31 (S)	0,000	0,000	0,000	-	0,000	0,000	-
H31(R)	0,000	0,000	0,000	-	0,000	0,000	-
G30	0,000	0,000	0,000	-	0,000	0,000	-
H32 (S)	0,000	0,000	0,000	-	0,000	0,000	-
H32(R)	0,000	0,000	0,000	-	0,000	0,000	-
H33 (S)	0,000	0,000	0,000	-	0,000	0,000	-
H33(R)	0,000	0,000	0,000	-	0,000	0,000	-
S20	0,249	0,234	0,237	-	0,240	0,008	3,3
S21	0,125	0,133	0,128	-	0,129	0,004	3,0
S22	0,106	0,099	0,105	-	0,103	0,004	3,6
D27(βs)	0,132	0,133	0,134	-	0,133	0,001	0,7
D27(βr)	0,052	0,056	0,054	-	0,054	0,002	3,6
D27 (as)	0,018	0,019	0,020	-	0,019	0,001	5,7
D27 (ar)	0,015	0,014	0,014	-	0,015	0,000	2,2
S27 (as)	0,020	0,020	0,017	-	0,019	0,001	7,8
S27 (βr)	0,073	0,081	0,078	-	0,078	0,004	5,0
S27 (βs)	0,068	0,077	0,070	-	0,072	0,005	6,4
S27 (ar)	0,017	0,016	0,015	-	0,016	0,001	5,6
S28 (as)	0,002	0,002	0,002	-	0,002	0,000	6,4
S28 (βr)	0,032	0,029	0,036	-	0,032	0,003	10,3
S28(βs)	0,009	0,008	0,008	-	0,008	0,000	5,5
S28 (ar)	0,023	0,020	0,022	-	0,021	0,001	6,7
S29 (as)	0,022	0,025	0,024	-	0,023	0,001	5,3
S29 (βr)	0,009	0,010	0,009	-	0,010	0,001	8,7
Si .(βs)	0,008	0,007	0,007	-	0,007	0,001	7,2
S29 (αr)	0,021	0,018	0,020	-	0,020	0,001	7,5
N	0,094	0,089	0,092	-	0,092	0,002	2,7
2-MN	0,104	0,109	0,109	-	0,107	0,003	2,8
1-MN	0,060	0,064	0,060	-	0,061	0,002	3,7
2-EN	0,040	0,040	0,045	-	0,042	0,003	7,2
1-EN	0,000	0,001	0,001	-	0,000	0,000	6,4
2.6 + 2.7 - DMN	0,104	0,110	0,109	-	0,108	0,003	3,1
1.3 + 1.7 - DMN	0,047	0,051	0,056	-	0,051	0,004	8,5
1,6 - DMN	0,030	0,034	0,032	-	0,032	0,002	5,7
1,4 + 2,3 - DMN	0,043	0,045	0,043	-	0,044	0,001	2,5
1,5 - DMN	0,009	0,008	0,010	-	0,009	0,001	6,7
1,2 - DMN	0,005	0,005	0,005	-	0,005	0,000	6,1
1,3,7 - TMN	0,041	0,044	0,041	-	0,042	0,002	3,9
1,3,6 - TMN	0,046	0,049	0,046	-	0,047	0,002	3,2
1,3,5 + 1,4,6 - TMN	0,007	0,007	0,007	-	0,007	0,000	6,0
2,3,6 - TMN	0,028	0,030	0,024	-	0,027	0,003	10,2
1,2,7 + 1,6,7 - TMN	0,000	0,000	0,000	-	0,000	0,000	-
1,2,6 - TMN	0,000	0,000	0,000	-	0,000	0,000	-
1,2,4 - TMN	0,001	0,001	0,001	-	0,001	0,000	6,8
1,2,5 - TMN	0,001	0,001	0,001	-	0,001	0,000	10,6
Ph	0,082	0,074	0,073	-	0,076	0,005	6,7

3-MPh	0,064	0,056	0,056	-	0,059	0,005	7,7
2-MPh	0,061	0,054	0,051	-	0,055	0,005	9,8
9-MPh	0,033	0,035	0,039	-	0,036	0,003	7,5
1-MPh	0,028	0,031	0,031	-	0,030	0,002	6,3
D	0,017	0,014	0,017	-	0,016	0,002	9,7
4-MDBT	0,033	0,029	0,030	-	0,031	0,002	7,3
2 + 3 - MDBT	0,021	0,020	0,021	-	0,021	0,001	3,1
1 - MDBT	0,001	0,001	0,001	-	0,001	0,000	7,1

Idem Table 11.

Table DB. Diagnostic ratios obtained from crude DB.

RDs	DB1	DB2	DB3	DB4	Prom	SDs	RSDs
P/F	1,612	1,651	1,630	-	1,631	0,019	1,2
P$\frac{1}{8}$-C$_{17}$	0,688	0,673	0,716	-	0,692	0,022	3,2
F$\frac{1}{8}$-C 8$_1$	0,469	0,447	0,479	-	0,465	0,016	3,5
n-C29$\frac{1}{8}$-Cj$_7$	0,271	0,237	0,232	-	0,246	0,021	8,6
H29/H30	-	-	-	-	-		-
10 x G30/G30 x C30	-	-	-	-	-		-
M30/H30	-	-	-	-	-		-
% S27	58,78	62,06	58,64	-	59,82	1,937	3,2
% S28	21,39	18,82	21,80	-	20,67	1,616	7,8
% S29	19,84	19,13	19,57	-	19,51	0,359	1,8
D27/S27	1,221	1,144	1,237	-	1,201	0,050	4,1
IMP	1,315	1,179	1,132	-	1,209	0,095	7,9
Rc	1,163	1,089	1,063	-	1,105	0,052	4,7
% 4-MeDBT	60,27	58,54	57,91	-	58,90	1,223	2,1
% 2+3-MeDBT	38,65	40,40	41,07	-	40,04	1,248	3,1
% 1-MeDBT	1,075	1,061	1,022	-	1,053	0,028	2,6
DBT/Ph	0,208	0,194	0,235	-	0,212	0,021	9,8

Idem Table 12.

Table EA. Relative standard deviation obtained from the relative abundances of the crude EA.

RAs	EA1	EA2	EA3	EA4	Averages	SDs	RSDs
n-C9	0,083	0,078	-	-	0,080	0,003	4,1
n-C10	0,098	0,094	-	-	0,096	0,003	2,9
n-C11	0,112	0,102	-	-	0,107	0,007	6,3
n-C12	0,084	0,081	-	-	0,083	0,002	2,8
n-C13	0,062	0,064	-	-	0,063	0,001	2,4
n-C14	0,051	0,052	-	-	0,052	0,001	1,6
n-C15	0,048	0,047	-	-	0,048	0,001	1,6
n-C16	0,040	0,041	-	-	0,041	0,001	2,2
n-C17	0,037	0,039	-	-	0,038	0,001	3,1
P	0,029	0,030	-	-	0,030	0,001	3,5
n-C18	0,031	0,034	-	-	0,033	0,002	7,3
F	0,012	0,012	-	-	0,012	0,000	1,0
n-C19	0,032	0,035	-	-	0,033	0,002	6,4
II-C20	0,030	0,033	-	-	0,031	0,002	7,2
n-C21	0,031	0,032	-	-	0,032	0,001	1,6
n-C22	0,032	0,032	-	-	0,032	0,000	0,4
n-C23	0,030	0,032	-	-	0,031	0,002	5,3
n-C24	0,030	0,032	-	-	0,031	0,001	4,7
n-C25	0,031	0,032	-	-	0,031	0,000	1,1
n-C26	0,028	0,026	-	-	0,027	0,001	4,1
n-C27	0,025	0,026	-	-	0,025	0,000	2,0
n-C28	0,020	0,020	-	-	0,020	0,000	0,5
n-C29	0,014	0,015	-	-	0,015	0,001	5,7
n-C30	0,010	0,009	-	-	0,010	0,001	6,7
T19	0,210	0,209	-	-	0,209	0,000	0,2
T20	0,040	0,038	-	-	0,039	0,001	2,9
T21	0,076	0,072	-	-	0,074	0,002	3,1
T23	0,082	0,070	-	-	0,076	0,008	10,6
T24	0,056	0,054	-	-	0,055	0,002	3,0
T25	0,000	0,000	-	-	0,000	0,000	-
T26 (R)	0,057	0,057	-	-	0,057	0,000	0,7
T26(S)	0,015	0,014	-	-	0,015	0,001	5,8
Ts	0,041	0,043	-	-	0,042	0,001	3,3
Tm	0,048	0,050	-	-	0,049	0,001	3,0
H29	0,131	0,133	-	-	0,132	0,002	1,2
H30	0,146	0,154	-	-	0,150	0,005	3,6
M30	0,012	0,012	-	-	0,012	0,000	4,0
H31(S)	0,022	0,025	-	-	0,024	0,002	9,0
H31(R)	0,016	0,019	-	-	0,017	0,002	10,3
G30	0,008	0,007	-	-	0,008	0,000	6,2

	EA1	EA2	EA3	EA4	Prom	SDs	RSDs
H_{32}(S)	0,015	0,016	-	-	0,015	0,001	5,9
H_{32}(R)	0,009	0,010	-	-	0,010	0,001	5,6
H_{33}(S)	0,008	0,009	-	-	0,009	0,001	7,6
H_{33}(R)	0,007	0,007	-	-	0,007	0,000	2,5
S_{20}	0,198	0,195	-	-	0,197	0,002	1,0
S_{21}	0,082	0,083	-	-	0,083	0,001	1,6
S_{22}	0,069	0,062	-	-	0,065	0,005	7,6
D_{27}(βs)	0,144	0,144	-	-	0,144	0,000	0,1
D_{27}(βr)	0,062	0,063	-	-	0,062	0,001	1,1
D_{27}(as)	0,032	0,033	-	-	0,033	0,001	1,9
D_{27}(ar)	0,022	0,022	-	-	0,022	0,000	1,2
S_{27}(as)	0,017	0,018	-	-	0,017	0,001	4,7
S_{27}(βr)	0,130	0,134	-	-	0,132	0,003	2,1
S_{27}(βs)	0,038	0,037	-	-	0,037	0,001	3,4
S_{27}(ar)	0,013	0,014	-	-	0,014	0,001	7,1
S_{28}(as)	0,016	0,015	-	-	0,015	0,001	6,1
S_{28}(βr)	0,028	0,030	-	-	0,029	0,001	3,9
S_{28}(βs)	0,019	0,019	-	-	0,019	0,000	1,6
S_{28}(ar)	0,027	0,026	-	-	0,027	0,000	0,7
S_{29}(as)	0,038	0,039	-	-	0,038	0,000	0,6
S_{29}(βr)	0,017	0,019	-	-	0,018	0,001	7,7
S_{29}(βs)	0,008	0,007	--		0,008	0,001	6,6
S_{29}(αr)	0,041	0,040	--		0,040	0,001	1,9
N	0,050	0,051	--		0,051	0,001	1,2
2-MN	0,112	0,102	--		0,107	0,007	6,4
1-MN	0,078	0,079	--		0,079	0,001	1,2
2-EN	0,021	0,022	--		0,021	0,001	3,3
1-EN	0,013	0,015	--		0,014	0,001	8,1
2.6 + 2.7 - DMN	0,080	0,083	--		0,082	0,002	2,5
1.3 + 1.7 - DMN	0,072	0,068	--		0,070	0,003	4,1
1,6 - DMN	0,047	0,050	--		0,048	0,002	4,1
1,4 + 2,3 - DMN	0,038	0,039	--		0,038	0,001	2,4
1,5 - DMN	0,014	0,016	--		0,015	0,001	7,8
1,2 - DMN	0,031	0,031	--		0,031	0,000	0,8
1,3,7 - TMN	0,049	0,051	--		0,050	0,001	2,3
1,3,6 - TMN	0,059	0,062	--		0,060	0,002	3,3
1,3,5 + 1,4,6 - TMN	0,043	0,046	--		0,045	0,002	4,8
2,3,6 - TMN	0,019	0,017	--		0,018	0,002	9,4
1,2,7 + 1,6,7 - TMN	0,022	0,023	--		0,022	0,001	3,0
1,2,6 - TMN	0,006	0,005	--		0,005	0,000	6,2
1,2,4 - TMN	0,011	0,011	--		0,011	0,000	0,4
1,2,5 - TMN	0,072	0,072	--		0,072	0,000	0,1
Ph	0,044	0,044	--		0,044	0,000	0,1
3-MPh	0,020	0,019	--		0,019	0,001	5,7
2-MPh	0,025	0,023	--		0,024	0,001	4,9
9-MPh	0,033	0,035	--		0,034	0,001	2,4
1-MPh	0,021	0,019	--		0,020	0,001	5,2
D	0,009	0,008	--		0,009	0,001	6,9
4-MDBT	0,007	0,006	--		0,007	0,000	7,1
2 + 3 - MDBT	0,003	0,003	--		0,003	0,000	5,9
1 - MDBT	0,002	0,001	--		0,002	0,000	1,3

Idem Table 11.

Table EA. Diagnostic ratios obtained from raw EA.

RDs	EA1	EA2	EA3	EA4	Prom	SDs	RSDs
P/F	2,392	2,550	-	-	2,471	0,112	4,5
P/n-C_{17}	0,777	0,782	-	-	0,780	0,004	0,5
F_{Zn}-Cx_8	0,389	0,346	-	-	0,367	0,031	8,3
n-C_{19}/n-C_{17}	0,375	0,389	-	-	0,382	0,010	2,6
H_{29}/H_{30}	0,894	0,864	-	-	0,879	0,021	2,4
$10 \times G_{30}/G_{30} \times C_{30}$	0,085	0,076	-	-	0,081	0,006	7,5
M_{30}/H_{30}	0,518	0,454	-	-	0,486	0,045	9,3
% S_{27}	50,55	51,01	-	-	50,78	0,327	0,6
% S_{28}	22,83	22,62	-	-	22,72	0,150	0,7
% S_{29}	26,62	26,37	-	-	26,50	0,178	0,7
D_{27}/S_{27}	1,309	1,291	-	-	1,300	0,013	1,0
IMP	0,690	0,643	-	-	0,666	0,033	5,0
Rc	0,819	0,794	-	-	0,806	0,018	2,3
% 4-MeDBT	58,72	57,79	-	-	58,26	0,656	1,1

% 2+3-MeDBT	28,35	28,39	-	-	28,37	0,034	0,1
% 1-MeDBT	12,93	13,81	-	-	13,37	0,622	4,7
DBT/Ph	0,205	0,186	-	-	0,195	0,013	6,9

Idem Table 12.

Table EB. Relative standard deviations obtained from the relative abundances of raw EB.

RAs	EB1	EB2	EB3	EB4	Averages	SDs	RSDs
n-C9	0,064	0,067	-	-	0,066	0,002	3,2
n-C10	0,089	0,088	-	-	0,089	0,001	0,7
n-C11	0,102	0,098	-	-	0,100	0,003	3,4
n-C12	0,091	0,089	-	-	0,090	0,001	1,3
n-Ci3	0,071	0,073	-	-	0,072	0,001	1,9
n-C14	0,058	0,060	-	-	0,059	0,001	2,1
n-C15	0,052	0,053	-	-	0,052	0,000	0,9
n-Ci6	0,045	0,045	-	-	0,045	0,000	0,4
n-C17	0,043	0,041	-	-	0,042	0,002	3,9
P	0,023	0,025	-	-	0,024	0,001	4,2
n-C18	0,035	0,036	-	-	0,035	0,001	1,4
F	0,009	0,009	-	-	0,009	0,000	3,2
n-C19	0,034	0,035	-	-	0,035	0,001	2,7
n-C20	0,033	0,033	-	-	0,033	0,000	0,9
n-C21	0,032	0,032	-	-	0,032	0,000	0,9
n-C22	0,031	0,031	-	-	0,031	0,000	0,3
n-C23	0,030	0,031	-	-	0,031	0,000	0,3
n-C24	0,031	0,030	-	-	0,030	0,000	0,7
n-C25	0,033	0,031	-	-	0,032	0,001	4,6
n-C26	0,025	0,025	-	-	0,025	0,000	0,2
n-C27	0,026	0,025	-	-	0,026	0,001	3,5
n-C28	0,018	0,019	-	-	0,019	0,001	6,8
n-C29	0,015	0,015	-	-	0,015	0,000	1,9
n-C30	0,010	0,009	-	-	0,009	0,001	6,1
T19	0,175	0,179	-	-	0,177	0,003	1,6
T20	0,032	0,030	-	-	0,031	0,002	5,1
T21	0,058	0,054	-	-	0,056	0,003	4,5
T23	0,074	0,079	-	-	0,077	0,004	4,9
T24	0,040	0,039	-	-	0,039	0,001	2,0
T25	0,000	0,000	-	-	0,000	0,000	-
T26 (R)	0,054	0,054	-	-	0,054	0,000	0,6
T26 (S)	0,013	0,012	-	-	0,012	0,000	2,3
Ts	0,036	0,037	-	-	0,037	0,000	0,9
Tm	0,057	0,058	-	-	0,058	0,001	1,4
H29	0,155	0,154	-	-	0,155	0,001	0,6
H30	0,193	0,187	-	-	0,190	0,004	2,1
M30	0,014	0,013	-	-	0,014	0,001	6,0
H31(S)	0,028	0,030	-	-	0,029	0,002	5,3
H31 (R)	0,018	0,020	-	-	0,019	0,001	6,0
G30	0,007	0,007	-	-	0,007	0,000	1,7
H32(S)	0,018	0,019	-	-	0,018	0,000	1,3
H32 (R)	0,010	0,011	-	-	0,011	0,000	3,6
H33(S)	0,010	0,010	-	-	0,010	0,000	4,8
H33 (R)	0,006	0,006	-	-	0,006	0,000	3,7
S_{20}	0,185	0,184	-	-	0,185	0,001	0,4
S21	0,081	0,083	-	-	0,082	0,001	1,2
S_{22}	0,064	0,062	-	-	0,063	0,002	2,7
D27(βs)	0,140	0,138	-	-	0,139	0,001	0,7
D27(βr)	0,061	0,062	-	-	0,061	0,001	0,9
D27 (as)	0,030	0,031	-	-	0,031	0,001	3,2
D27 (ar)	0,020	0,022	-	-	0,021	0,001	6,6
S27 (as)	0,021	0,024	-	-	0,023	0,002	7,6
S27(βr)	0,145	0,144	-	-	0,144	0,001	0,5
S27 (βs)	0,029	0,034	-	-	0,032	0,003	9,4
S27 (ar)	0,015	0,013	-	-	0,014	0,002	10,9
S28 (as)	0,020	0,018	-	-	0,019	0,001	5,9
S28(βr)	0,026	0,028	-	-	0,027	0,002	5,6
S28 (βs)	0,018	0,021	-	-	0,019	0,002	10,7
S28 (ar)	0,027	0,026	-	-	0,027	0,001	1,9
S29 (as)	0,043	0,041	-	-	0,042	0,001	3,5
S29(βr)	0,016	0,017	-	-	0,017	0,001	3,0
S29(βs)	0,010	0,010	--		0,010	0,000	3,2
S29 (ar)	0,047	0,042	--		0,045	0,004	7,9
N	0,058	0,051	--		0,054	0,005	9,3
2-MN	0,088	0,100	--		0,094	0,009	9,3
1-MN	0,069	0,079	--		0,074	0,006	8,7
2-EN	0,028	0,025	--		0,027	0,002	6,8
1-EN	0,017	0,016	--		0,016	0,001	4,4
2.6 + 2.7 - DMN	0,084	0,096	--		0,090	0,008	9,2

1.3 + 1.7 - DMN	0,067	0,063	--		0,065	0,003	3,9
1,6 - DMN	0,044	0,043	--		0,044	0,001	1,7
1,4 + 2,3 - DMN	0,042	0,044	--		0,043	0,001	3,5
1,5 - DMN	0,019	0,017	--		0,018	0,001	5,8
1,2 - DMN	0,038	0,033	--		0,036	0,004	10,4
1,3,7 - TMN	0,044	0,040	--		0,042	0,003	6,9
1,3,6 - TMN	0,053	0,052	--		0,053	0,000	0,9
1,3,5 + 1,4,6 - TMN	0,045	0,044	--		0,044	0,001	1,7
2,3,6 - TMN	0,020	0,020	--		0,020	0,000	1,9
1,2,7 + 1,6,7 - TMN	0,040	0,036	--		0,038	0,003	8,0
1,2,6 - TMN	0,013	0,012	--		0,012	0,001	6,4
1,2,4 - TMN	0,013	0,012	--		0,012	0,001	8,3
1,2,5 - TMN	0,058	0,067	--		0,063	0,007	11,0
Ph	0,043	0,041	--		0,042	0,001	3,1
3-MPh	0,020	0,019	--		0,019	0,001	5,6
2-MPh	0,026	0,023	--		0,025	0,002	6,7
9-MPh	0,034	0,032	--		0,033	0,001	3,1
1-MPh	0,021	0,018	--		0,020	0,002	7,8
D	0,004	0,005	--		0,005	0,000	6,4
4-MDBT	0,007	0,006	--		0,007	0,001	8,4
2 + 3 - MDBT	0,004	0,004	--		0,004	0,000	3,5
1 - MDBT	0,001	0,001	--		0,001	0,000	1,5

Idem Table 11.

Table EB. Diagnostic ratios obtained from raw EB.

RDs	EB1	EB2	EB3	EB4	Prom	SDs	RSDs
P/F	2,594	2,880	-	-	2,737	0,202	7,4
P/n-C17	0,541	0,606	-	-	0,574	0,046	8,1
F/n-C$_{18}$	0,258	0,242	-	-	0,250	0,012	4,6
n-C29%-C17	0,342	0,371	-	-	0,356	0,021	5,8
H29/H30	0,805	0,822	-	-	0,813	0,012	1,5
10 x G30/G30 X C30	0,074	0,070	-	-	0,072	0,003	3,9
M30/H30	0,344	0,346	-	-	0,345	0,001	0,3
% S$_{27}$	50,34	51,22	-	-	50,78	0,621	1,2
% S28	21,72	22,43	-	-	22,08	0,500	2,3
% S29	27,94	26,35	-	-	27,14	1,121	4,1
D27/S27	1,189	1,184	-	-	1,186	0,004	0,3
IMP	0,705	0,684	-	-	0,694	0,015	2,2
Rc	0,828	0,816	-	-	0,822	0,008	1,0
% 4-MeDBT	57,96	56,05	-	-	57,01	1,351	2,4
% 2+3-MeDBT	30,69	31,84	-	-	31,27	0,811	2,6
% 1-MeDBT	11,34	12,11	-	-	11,73	0,540	4,6
DBT/Ph	0,100	0,114	-	-	0,107	0,010	9,5

Idem Table 12.

Table FI. Relative standard deviation obtained from the relative abundances (RAs) of the FI crude oil.

RAs	FI1	FI2	FI3	FI4	Averages	SDs	RSDs
n-C9	0,057	0,061	0,057	-	0,058	0,002	3,8
n-C10	0,060	0,062	0,051	-	0,058	0,006	10,1
n-C11	0,062	0,061	0,054	-	0,059	0,004	7,1
n-C12	0,059	0,058	0,052	-	0,057	0,004	7,4
n-C13	0,055	0,054	0,056	-	0,055	0,001	1,9
n-C14	0,054	0,053	0,056	-	0,055	0,002	3,1
n-C15	0,056	0,055	0,056	-	0,056	0,001	1,4
n-C16	0,057	0,052	0,055	-	0,054	0,002	4,1
n-C17	0,049	0,052	0,055	-	0,052	0,003	5,8
P	0,034	0,034	0,036	-	0,035	0,002	4,3
n-C18	0,045	0,046	0,049	-	0,047	0,002	4,0
F	0,017	0,018	0,020	-	0,019	0,001	7,9
n-C19	0,046	0,047	0,049	-	0,047	0,001	2,9
n-C20	0,044	0,043	0,045	-	0,044	0,001	2,3
n-C21	0,042	0,044	0,045	-	0,044	0,002	3,9
n-C22	0,040	0,041	0,042	-	0,041	0,001	2,2
n-C23	0,039	0,039	0,039	-	0,039	0,000	0,4
n-C24	0,037	0,037	0,038	-	0,037	0,001	1,9
n-C25	0,035	0,035	0,037	-	0,036	0,001	2,2
n-C26	0,029	0,029	0,029	-	0,029	0,000	1,1
n-C27	0,026	0,026	0,026	-	0,026	0,000	1,6
n-C28	0,023	0,020	0,021	-	0,021	0,002	7,4
n-C29	0,020	0,018	0,018	-	0,019	0,001	6,1
n-C30	0,013	0,014	0,013	-	0,013	0,000	3,7
T19	0,159	0,145	0,144	-	0,149	0,008	5,6
T20	0,044	0,044	0,041	-	0,043	0,002	3,5
T2i	0,109	0,111	0,107	-	0,109	0,002	1,9

RDs	FI1	FI2	FI3	FI4	Prom	SDs	RSDs
T23	0,119	0,114	0,108	-	0,114	0,005	4,8
T24	0,083	0,085	0,083	-	0,084	0,001	1,1
T25	0,000	0,000	0,000	-	0,000	0,000	#DIV/0!
T26(R)	0,063	0,065	0,066	-	0,065	0,002	2,5
T26(S)	0,021	0,022	0,021	-	0,021	0,001	3,3
Ts	0,058	0,058	0,056	-	0,057	0,001	2,1
Tm	0,039	0,043	0,039	-	0,040	0,002	5,9
H29	0,066	0,069	0,073	-	0,069	0,004	5,1
H30	0,151	0,154	0,170	-	0,158	0,010	6,6
M30	0,012	0,013	0,014	-	0,013	0,001	8,2
H31(S)	0,021	0,019	0,019	-	0,019	0,001	5,5
H31(R)	0,016	0,015	0,016	-	0,016	0,001	4,6
G30	0,014	0,014	0,013	-	0,014	0,001	3,8
H32(S)	0,012	0,013	0,013	-	0,013	0,000	3,1
H32(R)	0,006	0,007	0,007	-	0,007	0,001	8,6
H33(S)	0,006	0,007	0,007	-	0,007	0,000	5,1
H33(R)	0,003	0,003	0,003	-	0,003	0,000	6,3
S20	0,197	0,190	0,182	-	0,190	0,007	3,9
S21	0,105	0,103	0,106	-	0,104	0,002	1,6
S22	0,082	0,080	0,083	-	0,082	0,002	2,2
D27(βs)	0,160	0,151	0,148	-	0,153	0,006	4,1
D27(βr)	0,059	0,060	0,056	-	0,058	0,002	3,8
D27(αs)	0,026	0,027	0,028	-	0,027	0,001	2,2
D27(ar)	0,018	0,020	0,021	-	0,020	0,001	6,8
S27(as)	0,024	0,024	0,023	-	0,024	0,001	2,8
S27(βr)	0,105	0,114	0,109	-	0,109	0,005	4,4
S27(βs)	0,044	0,051	0,051	-	0,049	0,004	8,9
S27(ar)	0,011	0,011	0,010	-	0,010	0,001	4,8
S28(as)	0,004	0,005	0,005	-	0,005	0,001	10,4
S28(βr)	0,033	0,034	0,035	-	0,034	0,001	2,8
S28(βs)	0,010	0,011	0,012	-	0,011	0,001	10,0
S28(ar)	0,026	0,029	0,030	-	0,028	0,002	7,4
S29(as)	0,035	0,034	0,039	-	0,036	0,003	7,1
S29(βr)	0,012	0,013	0,012	-	0,012	0,001	4,2
S29(βs)	0,008	0,008	0,008	-	0,008	0,000	1,5
S29(αr)	0,038	0,035	0,040	-	0,038	0,002	6,6
N	0,063	0,065	0,071	-	0,066	0,004	6,7
2-MN	0,085	0,084	0,080	-	0,083	0,003	3,0
1-MN	0,072	0,065	0,062	-	0,066	0,005	7,7
2-EN	0,022	0,023	0,020	-	0,021	0,002	7,2
1-EN	0,009	0,010	0,009	-	0,009	0,001	8,5
2.6 + 2.7 - DMN	0,101	0,103	0,118	-	0,107	0,009	8,6
1.3 + 1.7 - DMN	0,037	0,042	0,043	-	0,041	0,003	7,5
1,6 - DMN	0,042	0,047	0,042	-	0,044	0,003	5,8
1,4 + 2,3 - DMN	0,040	0,041	0,039	-	0,040	0,001	2,7
1,5 - DMN	0,013	0,015	0,015	-	0,014	0,001	8,6
1,2 - DMN	0,021	0,020	0,020	-	0,020	0,001	3,7
1,3,7 - TMN	0,048	0,044	0,048	-	0,047	0,002	4,9
1,3,6 - TMN	0,055	0,054	0,053	-	0,054	0,001	2,2
1,3,5 + 1,4,6 - TMN	0,023	0,021	0,021	-	0,022	0,001	4,8
2,3,6 - TMN	0,022	0,020	0,021	-	0,021	0,001	3,6
1,2,7 + 1,6,7 - TMN	0,016	0,019	0,018	-	0,018	0,001	7,1
1,2,6 - TMN	0,020	0,022	0,022	-	0,022	0,001	5,3
1,2,4 - TMN	0,008	0,010	0,009	-	0,009	0,001	8,4
1,2,5 - TMN	0,048	0,048	0,049	-	0,048	0,000	0,7
Ph	0,059	0,055	0,050	-	0,055	0,004	7,5
3-MPh	0,038	0,040	0,041	-	0,040	0,001	2,9
2-MPh	0,049	0,046	0,041	-	0,045	0,004	8,0
9-MPh	0,045	0,042	0,040	-	0,042	0,002	5,3
1-MPh	0,031	0,031	0,034	-	0,032	0,002	5,2
D	0,016	0,015	0,013	-	0,015	0,001	8,7
4-MDBT	0,010	0,011	0,012	-	0,011	0,001	6,6
2 + 3 - MDBT	0,006	0,006	0,007	-	0,007	0,001	11,4
1 - MDBT	0,002	0,002	0,002	-	0,002	0,000	1,1

Idem Table 11.

Table FI. Diagnostic ratios obtained from FI crude oil.

RDs	FI1	FI2	FI3	FI4	Prom	SDs	RSDs
P/F	1,937	1,858	1,799	-	1,865	0,070	3,7
PZn-C17	0,681	0,657	0,659	-	0,666	0,014	2,0

RAs	IMF	FM2	FM3	FM4	Averages	SDs	RSDs
$FZn\text{-}C_{18}$	0,387	0,397	0,418	-	0,401	0,016	3,9
$n\text{-}C_{2}$ %-C17	0,402	0,341	0,327	-	0,357	0,040	11,2
H29/H30	0,435	0,448	0,428	-	0,437	0,010	2,3
10 × G30G30 × C30	0,078	0,085	0,081	-	0,081	0,004	4,5
M30/H30	0,828	0,855	0,727	-	0,803	0,068	8,4
% S27	52,25	54,43	51,34	-	52,67	1,590	3,0
% S18	21,22	21,11	22,24	-	21,52	0,627	2,9
% S29	26,53	24,46	26,41	-	25,80	1,165	4,5
D_{27}/S_{27}	1,439	1,289	1,310	-	1,346	0,081	6,0
IMP	0,975	1,010	0,990	-	0,992	0,017	1,8
Rc	0,976	0,995	0,985	-	0,985	0,010	1,0
% 4-MeDBT	57,75	58,76	56,32	-	57,61	1,226	2,1
% 2+3-MeDBT	33,70	33,28	36,29	-	34,42	1,632	4,7
% 1-MeDBT	8,549	7,957	7,384	-	7,963	0,582	7,3
DBTZPh	0,273	0,275	0,266	-	0,271	0,004	1,6

Idem Table 12.

Table FM. Relative standard deviation obtained from the relative abundances of the raw FM.

RAs	IMF	FM2	FM3	FM4	Averages	SDs	RSDs
n-C9	0,002	0,002	-	-	0,002	0,000	1,5
n-C10	0,003	0,004	-	-	0,003	0,000	8,6
n-C11	0,006	0,007	-	-	0,007	0,001	8,3
n-C12	0,016	0,017	-	-	0,017	0,000	1,6
n-C13	0,030	0,031	-	-	0,031	0,001	2,4
n-C14	0,046	0,048	-	-	0,047	0,002	3,5
n-C15	0,061	0,061	-	-	0,061	0,000	0,2
n-C16	0,065	0,064	-	-	0,064	0,001	1,1
n-C17	0,068	0,070	-	-	0,069	0,001	1,6
P	0,045	0,046	-	-	0,046	0,001	1,6
n-C18	0,063	0,063	-	-	0,063	0,001	0,9
F	0,025	0,027	-	-	0,026	0,001	4,8
n-C19	0,066	0,063	-	-	0,064	0,002	3,0
n-C20	0,062	0,062	-	-	0,062	0,000	0,0
n-C21	0,061	0,061	-	-	0,061	0,000	0,5
n-C22	0,058	0,059	-	-	0,059	0,000	0,4
n-C23	0,056	0,057	-	-	0,056	0,001	1,9
n-C24	0,054	0,052	-	-	0,053	0,002	2,9
n-C25	0,053	0,047	-	-	0,050	0,004	7,1
n-C26	0,043	0,043	-	-	0,043	0,000	0,1
n-C27	0,042	0,039	-	-	0,041	0,002	5,4
n-C28	0,033	0,033	-	-	0,033	0,000	0,1
n-C29	0,025	0,028	-	-	0,027	0,002	6,3
n-C30	0,017	0,018	-	-	0,018	0,001	3,2
T19	0,107	0,108	-	-	0,107	0,000	0,3
T20	0,015	0,017	-	-	0,016	0,001	8,7
T2i	0,107	0,104	-	-	0,105	0,003	2,5
T23	0,143	0,142	-	-	0,142	0,001	0,5
T24	0,086	0,079	-	-	0,083	0,005	5,6
T25	0,000	0,000	-	-	0,000	0,000	-
T26(R)	0,073	0,075	-	-	0,074	0,001	1,0
T26(S)	0,028	0,029	-	-	0,029	0,001	3,0
Ts	0,066	0,064	-	-	0,065	0,002	2,4
Tm	0,049	0,053	-	-	0,051	0,003	5,6
H29	0,056	0,061	-	-	0,058	0,004	6,3
H30	0,156	0,149	-	-	0,153	0,005	3,4
M30	0,013	0,014	-	-	0,013	0,001	5,9
H31(S)	0,028	0,031	-	-	0,030	0,002	7,0
H31(R)	0,019	0,021	-	-	0,020	0,001	7,1
G30	0,011	0,011	-	-	0,011	0,000	1,0
H32(S)	0,016	0,015	-	-	0,016	0,000	1,8
H32(R)	0,011	0,012	-	-	0,011	0,001	6,7
H33(S)	0,009	0,010	-	-	0,010	0,000	1,2
H33(R)	0,007	0,006	-	-	0,006	0,000	2,0
S20	0,180	0,167	-	-	0,174	0,009	5,4
S21	0,089	0,092	-	-	0,090	0,003	2,9
S22	0,080	0,078	-	-	0,079	0,002	2,0
D27(βs)	0,142	0,144	-	-	0,143	0,002	1,3
D27(βr)	0,060	0,060	-	-	0,060	0,000	0,8
D27(αs)	0,026	0,025	-	-	0,025	0,001	2,9
D27(ar)	0,055	0,058	-	-	0,057	0,002	3,3
S27(as)	0,011	0,011	-	-	0,011	0,000	1,9
S27(βr)	0,116	0,113	-	-	0,115	0,002	1,5
S27(βs)	0,056	0,052	-	-	0,054	0,003	5,1

s_{27}(ar)	0,010	0,011	-	-	0,011	0,000	3,8
s_{28}(as)	0,012	0,012	-	-	0,012	0,000	2,9
s_{28}(βr)	0,047	0,046	-	-	0,047	0,000	1,0
s_{28}(βs)	0,010	0,011	-	-	0,010	0,001	8,5
s_{28}(ar)	0,020	0,022	-	-	0,021	0,002	8,3
s_{29}(as)	0,034	0,039	-	-	0,037	0,003	9,3
s_{29}(βr)	0,016	0,017	-	-	0,016	0,001	4,2
$S_2 9$(βs)	0,007	0,006	--		0,006	0,000	4,0
s_{29}(αr)	0,031	0,034	--		0,032	0,002	7,5
N	0,057	0,063	--		0,060	0,004	7,1
2-MN	0,062	0,066	--		0,064	0,003	5,0
1-MN	0,049	0,053	--		0,051	0,003	5,1
2-EN	0,026	0,027	--		0,026	0,001	3,1
1-EN	0,010	0,009	--		0,009	0,000	5,0
2.6 + 2.7 - DMN	0,066	0,067	--		0,067	0,001	1,0
1.3 + 1.7 - DMN	0,060	0,059	--		0,060	0,000	0,2
1,6 - DMN	0,045	0,047	--		0,046	0,001	2,1
1,4 + 2,3 - DMN	0,027	0,029	--		0,028	0,001	4,5
1,5 - DMN	0,015	0,015	--		0,015	0,000	1,0
1,2 - DMN	0,028	0,028	--		0,028	0,000	1,4
1,3,7 - TMN	0,029	0,029	--		0,029	0,000	0,1
1,3,6 - TMN	0,062	0,071	--		0,066	0,006	9,0
1,3,5 + 1,4,6 - TMN	0,046	0,043	--		0,044	0,002	4,5
2,3,6 - TMN	0,031	0,030	--		0,031	0,001	2,9
1,2,7 + 1,6,7 - TMN	0,044	0,045	--		0,044	0,000	0,7
1,2,6 - TMN	0,033	0,032	--		0,033	0,001	3,4
1,2,4 - TMN	0,009	0,008	--		0,008	0,000	4,9
1,2,5 - TMN	0,044	0,040	--		0,042	0,003	7,0
Ph	0,059	0,057	--		0,058	0,002	2,8
3-MPh	0,038	0,036	--		0,037	0,001	4,0
2-MPh	0,048	0,042	--		0,045	0,004	9,6
9-MPh	0,043	0,038	--		0,040	0,004	10,1
1-MPh	0,030	0,031	--		0,031	0,001	2,2
D	0,017	0,016	--		0,016	0,001	4,8
4-MDBT	0,012	0,012	--		0,012	0,000	1,7
2 + 3 - MDBT	0,008	0,007	--		0,008	0,001	7,2
1 - MDBT	0,001	0,001	--		0,001	0,000	0,4

Idem Table 11.

Table FM. Diagnostic ratios obtained from the raw FM.

RDs	FM1	FM2	FM3	FM4	Prom	SDs	RSDs
P/F	1,804	1,727	-	-	1,766	0,055	3,1
PZn-C$_{17}$	0,666	0,667	-	-	0,666	0,000	0,0
FZn-C$_{18}$	0,396	0,429	-	-	0,413	0,023	5,7
n-C$_2$ 9⅛-C$_{17}$	0,373	0,398	-	-	0,385	0,018	4,7
H29/H30	0,357	0,410	-	-	0,384	0,037	9,7
10 x G$_{30}$G$_{30}$ x C$_{30}$	0,082	0,093	-	-	0,087	0,008	9,2
M30/H30	0,655	0,694	-	-	0,675	0,028	4,1
% S$_{27}$	52,40	49,98	-	-	51,19	1,710	3,3
% S$_{18}$	23,90	24,39	-	-	24,15	0,346	1,4
% S$_{29}$	23,70	25,62	-	-	24,66	1,364	5,5
D_{27}/S_{27}	1,464	1,532	-	-	1,498	0,049	3,2
IMP	0,962	0,920	-	-	0,941	0,030	3,2
Rc	0,969	0,946	-	-	0,958	0,016	1,7
% 4-MeDBT	56,43	59,17	-	-	57,80	1,936	3,3
% 2+3-MeDBT	37,56	34,71	-	-	36,14	2,011	5,6
% 1-MeDBT	6,012	6,119	-	-	6,066	0,075	1,2
DBTZPh	0,284	0,276	-	-	0,280	0,006	2,0

Idem Table 12.

Table FS. Relative standard deviation obtained from the relative abundances of the raw FS.

RAs	FS1	FS2	FS3	FS4	Averages	SDs	RSDs
n-C9	0,046	0,048	-	-	0,047	0,001	2,5
n-C10	0,058	0,062	-	-	0,060	0,003	5,2
n-C11	0,069	0,065	-	-	0,067	0,003	3,9
n-C12	0,059	0,052	-	-	0,056	0,005	8,6
n-C13	0,057	0,055	-	-	0,056	0,001	2,0
n-C14	0,054	0,053	-	-	0,053	0,001	1,5
n-C15	0,055	0,059	-	-	0,057	0,003	5,3

n-C16	0,055	0,055	-	-	0,055	0,000	0,3
n-C17	0,051	0,055	-	-	0,053	0,003	5,0
P	0,035	0,038	-	-	0,037	0,002	5,5
n-C18	0,048	0,049	-	-	0,048	0,001	1,3
F	0,018	0,020	-	-	0,019	0,001	5,6
n-C19	0,047	0,049	-	-	0,048	0,001	3,1
n-C20	0,044	0,044	-	-	0,044	0,000	0,8
n-C21	0,041	0,046	-	-	0,043	0,004	9,0
n-C22	0,045	0,041	-	-	0,043	0,003	6,3
n-C23	0,040	0,037	-	-	0,038	0,002	4,0
n-C24	0,038	0,037	-	-	0,038	0,001	1,5
n-C25	0,035	0,034	-	-	0,034	0,000	0,5
n-C26	0,030	0,028	-	-	0,029	0,001	3,3
n-C27	0,027	0,024	-	-	0,025	0,002	8,4
n-C28	0,023	0,022	-	-	0,023	0,001	2,9
n-C29	0,017	0,016	-	-	0,016	0,000	2,6
n-C30	0,009	0,009	-	-	0,009	0,000	3,4
T19	0,156	0,153	-	-	0,154	0,002	1,3
T20	0,040	0,038	-	-	0,039	0,001	3,4
T21	0,106	0,099	-	-	0,103	0,005	5,1
T23	0,110	0,105	-	-	0,107	0,004	3,3
T24	0,077	0,070	-	-	0,073	0,005	6,6
T25	0,000	0,000	-	-	0,000	0,000	-
T26(R)	0,058	0,053	-	-	0,056	0,004	6,7
T26(S)	0,020	0,023	-	-	0,021	0,003	12,1
Ts	0,052	0,058	-	-	0,055	0,004	6,8
Tm	0,037	0,042	-	-	0,040	0,004	9,2
H29	0,056	0,063	-	-	0,059	0,005	8,1
H30	0,169	0,171	-	-	0,170	0,002	1,0
M30	0,014	0,015	-	-	0,015	0,001	4,9
H31(S)	0,026	0,027	-	-	0,027	0,001	3,9
H31(R)	0,018	0,019	-	-	0,019	0,000	2,2
G30	0,012	0,012	-	-	0,012	0,000	2,3
H32(S)	0,015	0,016	-	-	0,016	0,001	3,5
H32(R)	0,014	0,015	-	-	0,014	0,001	6,4
H33(S)	0,010	0,011	-	-	0,011	0,000	4,3
H33(R)	0,010	0,010	-	-	0,010	0,000	3,1
S20	0,197	0,192	-	-	0,194	0,004	1,8
S21	0,090	0,084	-	-	0,087	0,004	4,5
S22	0,083	0,090	-	-	0,087	0,005	5,8
D27 (βs)	0,137	0,131	-	-	0,134	0,004	3,2
D27 (βr)	0,056	0,054	-	-	0,055	0,001	2,7
D27 (as)	0,023	0,023	-	-	0,023	0,000	0,8
D27 (ar)	0,051	0,059	-	-	0,055	0,005	9,6
S27 (as)	0,009	0,010	-	-	0,009	0,001	8,4
S27 (βr)	0,098	0,107	-	-	0,103	0,006	6,3
S27 (βs)	0,054	0,049	-	-	0,052	0,003	6,7
S27 (ar)	0,012	0,013	-	-	0,012	0,000	2,2
S28 (as)	0,011	0,012	-	-	0,011	0,001	5,7
S28 (βr)	0,036	0,035	-	-	0,035	0,001	2,3
S28 (βs)	0,013	0,012	-	-	0,013	0,001	7,4
S28 (ar)	0,032	0,029	-	-	0,031	0,002	5,7
S29 (as)	0,033	0,035	-	-	0,034	0,001	2,8
S29 (βr)	0,016	0,017	-	-	0,017	0,001	4,5
S29 (βs)	0,010	0,010	-	-	0,010	0,000	4,2
S29 (αr)	0,037	0,038	-	-	0,038	0,001	2,0
N	0,089	0,085	-	-	0,087	0,003	3,4
2-MN	0,088	0,091	-	-	0,090	0,002	2,4
1-MN	0,068	0,072	-	-	0,070	0,003	4,2
2-EN	0,028	0,029	-	-	0,029	0,000	0,2
1-EN	0,009	0,010	-	-	0,010	0,001	8,2
2.6 + 2.7 - DMN	0,089	0,090	-	-	0,090	0,000	0,3
1.3 + 1.7 - DMN	0,034	0,037	-	-	0,036	0,002	5,5
1,6 - DMN	0,025	0,026	-	-	0,025	0,001	3,6
1,4 + 2,3 - DMN	0,045	0,046	-	-	0,045	0,001	2,5
1,5 - DMN	0,015	0,014	-	-	0,015	0,001	5,5
1,2 - DMN	0,024	0,027	-	-	0,026	0,002	6,7
1,3,7 - TMN	0,045	0,045	-	-	0,045	0,000	0,4
1,3,6 - TMN	0,055	0,055	-	-	0,055	0,000	0,1
1,3,5 + 1,4,6 - TMN	0,025	0,027	-	-	0,026	0,002	6,9
2,3,6 - TMN	0,023	0,021	-	-	0,022	0,002	7,9
1,2,7 + 1,6,7 - TMN	0,021	0,022	-	-	0,022	0,001	5,7
1,2,6 - TMN	0,004	0,003	-	-	0,003	0,000	7,4

1,2,4 - TMN	0,009	0,008	-	-	0,009	0,001	8,6
1,2,5 - TMN	0,052	0,052	-	-	0,052	0,000	0,9
Ph	0,058	0,057	-	-	0,058	0,001	1,2
3-MPh	0,041	0,039	-	-	0,040	0,001	3,6
2-MPh	0,050	0,047	-	-	0,048	0,002	3,5
9-MPh	0,043	0,042	-	-	0,042	0,001	2,5
1-MPh	0,034	0,030	-	-	0,032	0,003	8,2
D	0,005	0,005	-	-	0,005	0,000	3,6
4-MDBT	0,012	0,010	-	-	0,011	0,001	8,8
2 + 3 - MDBT	0,007	0,006	-	-	0,007	0,001	7,9
1 - MDBT	0,002	0,002	-	-	0,002	0,000	0,2

Idem Table 11.

Table FS. Diagnostic ratios obtained from crude oil FS.

ii	FS1	FS2	FS3	FS4	Prom	SDs	RSDs
P/F	1,943	1,940	-	-	1,941	0,003	0,1
P/n-C17	0,697	0,703	-	-	0,700	0,004	0,5
FZn-Cx8	0,381	0,405	-	-	0,393	0,017	4,3
n-C19/n-C17	0,326	0,293	-	-	0,310	0,024	7,6
H29/H30	0,330	0,365	-	-	0,347	0,024	7,0
10 x G30/G30 X C30	0,085	0,090	-	-	0,087	0,003	3,8
M30/H30	0,682	0,653	-	-	0,668	0,021	3,1
% S27	47,77	48,73	-	-	48,25	0,681	1,4
% S28	25,38	23,96	-	-	24,67	0,998	4,0
% S29	26,86	27,31	-	-	27,08	0,317	1,2
D27/S27	1,551	1,496	-	-	1,523	0,039	2,5
IMP	1,004	1,000	-	-	1,002	0,003	0,3
Rc	0,992	0,990	-	-	0,991	0,002	0,2
% 4-MeDBT	57,65	56,77	-	-	57,21	0,627	1,1
% 2+3-MeDBT	34,11	34,02	-	-	34,06	0,063	0,2
% 1-MeDBT	8,241	9,217	-	-	8,729	0,690	7,9
DBT/Ph	0,086	0,083	-	-	0,084	0,002	2,4

Idem Table 12.

I want morebooks!

Buy your books fast and straightforward online - at one of world's fastest growing online book stores! Environmentally sound due to Print-on-Demand technologies.

Buy your books online at
www.morebooks.shop

Kaufen Sie Ihre Bücher schnell und unkompliziert online – auf einer der am schnellsten wachsenden Buchhandelsplattformen weltweit! Dank Print-On-Demand umwelt- und ressourcenschonend produzi ert.

Bücher schneller online kaufen
www.morebooks.shop

Printed by Books on Demand GmbH, Norderstedt / Germany